英国赫特福德大学服装设计硕士研究生毕业设计范本

第六届创意中国·全国设计艺术大奖赛 服装设计类 一等奖

空间 的 遐想
——将雕塑艺术导入时装设计
The Imagination of Space
--Tansforming Sculpture into Fashion

徐越 著
Yue Xu

西泠印社出版社

图书在版编目（CIP）数据

空间的遐想：将雕塑艺术导入时装设计 / 徐越著
. -- 杭州：西泠印社出版社，2013.3
ISBN 978-7-5508-0709-9

Ⅰ. ①空… Ⅱ. ①徐… Ⅲ. ①服装设计 Ⅳ. ①TS941.2

中国版本图书馆CIP数据核字(2013)第056188号

本书提要

本作品的灵感源来自于抽象艺术。设计创作的构思点基于英国著名抽象雕塑艺术大师芭芭拉·赫普沃斯女士的作品研究，以及对其他具有相同艺术思想及作品的研究借鉴之上，构思的重点是服装结构实与虚的平衡关系，探索的要点是建立人体曲线与服装廓形之间的空间美，寻求突破的是传统时尚观念与现代审美意识的自然与抽象。作品效果舒展流畅、端庄典雅、灵动美妙，整个设计理念及制作过程、思维开放，步骤清晰，结构严谨，新颖独创，成功地拓展了时装设计的空间领域。

空间的遐想——将雕塑艺术导入服装设计

徐越 著

特邀编辑：徐经验

责任编辑：王欣、郑晨

责任出版：李兵

装帧设计：徐越

出版发行：西泠印社出版社

地址：杭州西湖文化广场32号E区5楼

经销：全国新华书店

印刷：浙江海虹彩色印务有限公司

开本：889×1194 1/16

印张：9.25

印数：0001-2000.00

版次：2013年3月第1版 第1次印刷

书号：ISBN 978-7-5508-0709-9

定价：98.00元

作者简介
About Author

徐越，

女，

双子座，

生于 1987 年 6 月

籍贯：浙江省松阳县

浙江理工大学服装设计专业文学学士

英国赫特福德大学服装设计专业艺术硕士

中国设计师协会 会员

2005 年 9 月，考入浙江科技学院服装设计专业。

2007 年，参加全省"2+2"大学在校本科优秀生选拔考试，以全省第二名的成绩考入浙江理工大学服装学院。在校期间，成绩优秀，多次获得奖学金。是班级和学生会干部。

2009 年，参加北京国际时装周的汇展。设计制作的"80/90 的彩虹童年"针织服饰被多家媒体争相报道，并作为三件学生优秀作品之一刊登在新浪网首页。

2009 年 6 月，毕业于浙江理工大学，文学学士。

2009 年 8 月，进入浙江纺织服装科技有限公司（原中国纺织总会服装技术开发中心）就业，分配工作为科研。

2010 年 9 月，进入英国赫特福德大学艺术与设计学院攻读服装设计硕士学位。在校期间，频获导师及专业教授的高度评价。并以"distinction"全学院第一的成绩毕业，获艺术硕士的学位。毕业作品及论文被英所读大学作为研究生辅助教材及范本。

2011 年 9 月，研究生毕业作品"空间的遐想"在英国 Art and Design Gallery 陈列展示

2011 年 11 月，国际顶尖服装品牌公司 --- 荷兰 Iris Van Herpen 邀请加入设计团队，因工作签证原因未能赴荷。

2012 年 1 月，任杭州纵邺（香港）服饰有限公司服装设计总监助理，协助女装品牌的开发。

2012 年 8 月，任全国知名女装企业卓尚服饰（杭州）有限公司（原浙江三彩服饰有限公司）品牌艺术总监助理。

2012 年 9 月，获中国设计师协会举办的创意中国·第六届全国设计艺术大奖赛服装设计一等奖第一名。作品编入《中国创意设计年鉴·2012》。

2012 年 10 月，成为中国设计师协会会员。

目录
Contents

P003 作者简介 /About Author

P007 英国导师的评价及推荐（代序）/ Recommendation Letter

P011 作品简介 /Design Project Introduction

P015 设计流程图 /Flow Chart

P017 灵感研究 /Research Inspiration

P018 赫普沃斯抽象雕塑研究 /Research into Barbara Hepworth

P030 相关设计师作品概念研究 /Research into Fashion Designers

P042 流行趋势研究 / Fashion Trend Research

P044 设计构思 / Design Process

P045 结构设计与面料处理实验 /Experiments with Structure Design and Fabric

P055 灵感和色彩主题板 / Concept and Colour Board

P060 服装系列设计拓展 /Design Development

P065 面料的选择与搭配 / Fabric Choices and Selection

P068 样衣制作及发展流程 / Prototyping Development

P079 款式结构图 /Flat Drawing

P084　　服装成衣搭配决策 /Decision Making

P090　　时装插画 / Fashion Illustration

P097　　时装摄影造型设计 /Styling Design and Selection

P104　　时装摄影 /Fashion Shoot

P120　　对未来研究的展望 /My Agenda for The Future

P125　　论文：空间的遐想 ——将雕塑艺术导入服装设计

　　　　　　Essay： The Imagination of Space ---Transforming Sculpture into Fashion

P137　　参考文献 / Reference

P138　　参阅书目 / Bibliography

P141　　附件一 / Attachment one

　　　　　　读研毕业作品在英国 Art and Design Gallery 展出

P142　　附件二 / Attachment two

　　　　　　Iris Van Herpen 公司简介及邀请加入设计团队函件

P144　　附件三 /Attachment three

　　　　　　《空间的遐想》获 2012 年第六届创意中国·全国设计艺术大奖赛 服装设计类 一等奖

英国导师的评价及推荐·代序

代序 I

硕士研究生院院长的评价及推荐
Reference from the Programme Leader

I am the Programme Leader for the MA Art and Design, at the University of Hertfordshire, England. I have known Yue Xu for 2 years; as a student on the MA Fashion Design and since then she has kept me updated about her progress in China.

Yue was an excellent student. She was fully committed to her studies and determined to make the most of her time at our University here in England. At the beginning of the MA Programme, Yue worked hard to improve her English and always persevered to express herself clearly and well. She made positive contributions to all seminars and class discussions, never afraid to ask for clarification if anything was unclear.

Yue gained experience of working directly with students from other disciplines within Design, as well as working alongside students from the wider postgraduate community within the School of Creative Arts. In support of her specialist study, Fashion Design, Yue worked with a small group of students under the guidance of a specialist tutor, with whom she worked closely to develop and refine a body of design work for her final MA assessment. As well as producing a collection of very individual garments, Yue documented her MA study extremely thoroughly, producing a book of professional quality to visualise all aspects of her design development. We will be using her book with students in the future, as an excellent example of the high standard we hope our MA students will achieve.

Yue's time in England gave her an understanding of the European design process and the global fashion market. She made many friends amongst both the staff and her fellow students. During her time with us, she was an excellent advocate for the potential of international study. She worked hard, with positive energy and enthusiasm. We would have been delighted had she decided to stay in England for further study and hope that we can collaborate with her in some way in the future.

Yue's experience here in England gives a very strong global perspective to her fashion design potential, and the knowledge and rigor of her research and design development would be a very constructive example to fashion students. I wish her every success.

With kind regards

Sally Freshwater:
MA RCA, BA(Hons) Goldsmiths, University of London.
Programme Leader MA Art & Design.
School of Creative Arts,
Faculty of Science, Technology and Creative Arts
University of Hertfordshire

我是英国赫特福德大学艺术与设计硕士研究生院院长，从徐越作为时装设计硕士研究生认识她开始已经有两年，她毕业回中国后一直和我保持联系。

徐越是一个非常出色的学生，在英国求学期间，充分利用一切机会勤奋学习。在硕士课程初期，她努力学习提高英语水平，使自己表达清晰、贴切。她积极参与研讨会和班级讨论，对任何不明白的问题都不会放过。

徐越和设计专业以及创意艺术学院的其他研究生一起紧密合作，获得了宝贵的经验。为完成硕士学位的毕业设计，她和几个同学一起在一个专业导师的指导下，专注于自己的服装设计专业，不但出色完成了毕业设计还另外设计了许多服装。不仅如此，徐越还将设计过程详细记录了下来，创作了一本有很高专业质量的画册，来展现她的作品的创意和制作过程。我们将把她的毕业作品集作为一个优秀硕士生能达到的高标准模板及范本来要求将来所有的硕士生。

在英国学习期间，徐越对欧洲的设计过程和全球的时装市场有了较深的了解，并和许多教师和同学结下了深厚友谊。她在我校学习期间的表现是所有留学生学习的榜样：她学习刻苦、积极上进、充满热情。我们很希望她当时能留在英国继续深造，也希望将来能和她在专业上有合作的机会。

徐越在英国的学习经历赋予她的时装设计以全球视角，而她的研究、设计和知识将会是学生学习的典范。

祝她成功！

此致

Sally Freshwater
赫特福德大学艺术与设计学院硕士研究生院院长、英国核心纺织艺术家、英国刺绣协会成员、英国国家工艺理事会成员、皇家艺术学院硕士、伦敦大学金匠学院优等学士、

地址：英国赫特福德大学创意艺术学院

代序 Ⅱ

时尚设计系主任的评价及推荐
Reference from MA Design Co-ordinator

我认识徐越已经两年了，第一年是我在英国赫特福德大学担任时装设计系主任期间，后面一年是2011年徐越从赫特福德大学毕业获得硕士学位（优异生）回到中国后 她一直和我保持联系。赫特福德大学的硕士学位课程是建立在享有国际声誉的本科时装课程基础之上的，硕士阶段在增强学生专业服装设计核心价值意识的同时，使他们形成自己独特的设计风格。学生们必须和创意产业其它专业（比如音乐和影视）的学生一起合作，在一年里必须致力于一个创意主题，直到完成达到专业高标准的毕业作品为止。

徐越不仅工作勤奋，而且具备不畏困难、勇于挑战的能力。她读研期间非常重视学习，获得了全年级最高成绩，对此，专业的相关老师并不感到惊讶。她努力工作，认真思考，既熟悉时装设计的前因后果，又能用娴熟的服装制作技术，设计和制作出令人兴奋的高品质成果。对细节和设计的专注贯穿她作品的各个方面，她创作的一本记录她硕士阶段学习过程以及服装设计创意的书记录这一切。在英国一年的学习加强了她对全球时装产业以及英国设计行业内的设计过程的理解。

徐越的作品反映了她的个性，那就是走进一间她所在的工作室，里面总会有灿烂的笑容，有时是愉悦的的笑声，但她永远不失认真与深度。她设计的服装只不过是冰山一角，只有在作品之外的细节里，你才会发现思想和技能的交融是她对项目的乐趣所在。徐越非常注意细节的处理，在设计上从不知足。她与老师和同学都能融洽相处。小组讨论时她积极提出自己的观点，同样也乐于接受积极的批评。

徐越谈到的几个理念，我们对此很有兴趣，因为设计是一个全球性的业务，分享技术、交流对设计和消费受众的理解，对学设计的学生都有颇有裨益。我觉得徐越有这样的意愿和能力。

基于上述理由，我很高兴推荐徐越，并祝她成功。

Julian Lindley：
时尚设计系主任、皇家艺术协会会员、特许设计师协会会员、英国高等教育学院研究员、

地址：英国赫特福德大学创意艺术学院

I have known Yue Xu for two years. The first year was as the MA Design Co-ordinator at the University of Hertfordshire from which Yue Xu graduated with an MA Fashion (Distinction) in 2011. Since Yue Xu's return to China she has kept me informed on how her career is unfolding.

Yue Xu proved herself to be both dedicated hard working and capable of taking on new challenges regardless of difficulty. MA study was very valuable to her and none of the staff involved with the Programme were surprised that she received one of the highest marks in the year. She is thoughtful about her work working hard to both understand the context of Fashion Design and delivering exciting outcomes of very high quality in the skills by which garments were manufactured. This attention to design and detail was carried through into all aspects of her work including a book which summerised her MA journey and motivations as a Fashion Designer. Studying in England for a year has strengthened her understanding of the global Fashion Industry as well as gaining an understanding of the design processes which are inherent within the UK Design Profession.

Yue Xu's work reflects her character, that is walking into a studio in her presence there is always a smile, sometimes a giggle, and always a seriousness and depth. The garments are just the tip of the iceberg and it is in the supporting work where you discover the interlinking of ideas to technical skill that the joy of the projects lie. There is attention to detail, nothing is ever good enough, Yue Xu always wants to redo work to improve. She got on well with both staff and contemporaries and was capable of putting her point of view across in group discussions as well as taking positive criticism.

In Yue Xu's communication for the post she speaks of several ideas, this is something of great interest to us as Design is a global business and sharing of skills as well as differing (and shared) understanding of both design and consumer audiences would be of value to students who study fashion. Yue Xu has the motivation to make things happen.

For the above reasons it is a pleasure to endorse Yue Xu's book, and I wish her every success in her application.

With kind regards

Julian Lindley
FRSA, MCSD, FHEA, MA Design Co-ordinator
School of Creative Arts,
Faculty of Science, Technology and Creative Arts
University of Hertfordshire

代序 III

硕士研究生导师的评价及推荐
Reference from MA Fashion Design Tutor

I worked with Yue Xu for a year as her MA Fashion Design Tutor.

Yue Xu entered the course with flat pattern cutting skills and good design skills. Throughout the course all challenges were approached with enthusiasm, she relished new ways of working and new ways of thinking about the design process and its development. The research undertaken and its development were approached with an open mind and sensitivity. Her commitment to her studies was constant and she approached all tasks with skill and imagination. It is hard to fault her process of design and subsequent problem solving.

Yue Xu has a strong aesthetic sensibility and used this throughout her work from inception to presentation enabling her to describe sometimes complex ideas. She responds well to criticism/instruction and has a refined ability to self direct, self criticise and make decisions. Yue Xu was active in group sessions always willing to share opinions and encourage her peers.

Yue Xu has very good interpersonal skills and was popular with both other students and staff.
Her organisation and time management is exceptional.

I am delighted that Yue Xu is an enthusiastic and creative individual well able to inform and inspire others.

I wish her every success and would look forward to the possibility of working with her again in the future.

Annie Shellard

Annie Shellard
Senior Lecturer
BA Hons Fashion, MA Design

我作为徐越的服装设计硕士导师，和她有一年的密切交往。

徐越一进入硕士课程就具备了良好的平面裁剪切割技能和设计技巧。在课程学习过程中，徐越以极大的热情迎接各种挑战，她总是以新的思考和工作方式进行创意和设计。她以虚心的态度从事研究，对学业的投入是持之以恒的，以技巧和想象力完成各项任务，很难在她的设计过程和随后的问题解决方案中发现错误。

徐越具有很强的审美感，这使得她能在作品的开始到呈现的各个阶段都能描述复杂的理念。她对批评指正能从容应对，有很好的自我管理、自我批评及决策能力。徐越在小组讨论时表现活跃，愿意与大家分享自己的观点并激励同学。

徐越人际交往能力好，在师生中都很受欢迎。

她的组织和时间管理能力也非常突出。

徐越的实践技能和思辨能力都非常强。她热情洋溢、富有创造力，能很好地引导和鼓舞其他人。

我祝她成功，并期望将来能有机会再和她一起工作。

Annie Shellard
高级讲师，服装专业优等学士，设计硕士
地址：英国赫特福德大学创意艺术学院

作品简介
Design Project
Introduction

实与虚空间结构的平衡

The balance of solid structure and the area those were not solid

本人的硕士研究生毕业时装设计作品重在于对实与虚关系的探讨，是对服装空间的一次大胆的遐想，也是一次非常有趣的探索性旅程。

作品的设计深受英国著名抽象艺术大师赫普沃斯女士艺术思想及其雕塑作品的启发（Hepworth，1956），人体与服装的空间关系是研究探讨的主题。设计构思的重点是服装结构实与虚的平衡关系，作品效果舒展流畅，端庄典雅，灵动美妙。探索过程思维开放，步骤清晰，结构严谨，新颖创新。

总之，该系列作品的整体设计思路是探讨人体与服装的空间关系。设计的创作构思是基于对赫普沃斯女士抽象雕塑的研究，以及对其他具有相同理念的设计师作品的探索及借鉴之上的。主旨是将抽象雕塑艺术导入服装设计，以拓展时装设计的空间领域。

本作品设计完成后获所有导师好评，以"distinction"第一名的优异成绩毕业，获服装设计艺术硕士的学位。并于2011年9月，在Art and Design Gallery陈列展示一周。硕士研究生毕业作品及论文被英所读大学作为研究生辅助教材及范本。

The project of my fashion design concentrates on solid shape and negative spaces. In this project, I designed a series of the ready-to-wear garment that explore the space between the human body and garment, this idea of which was influenced by Barbara Hepworth's abstract sculpture (Hepworth, 1956). In particularly, I focused on the balance of solid structure and the area that were not solid, in order to create the garment to have clear silhouette and elegant detail.

Therefore, this section will present the ideas that my design was based on Barbara Hepworth's abstract sculpture and other designers' ideas.

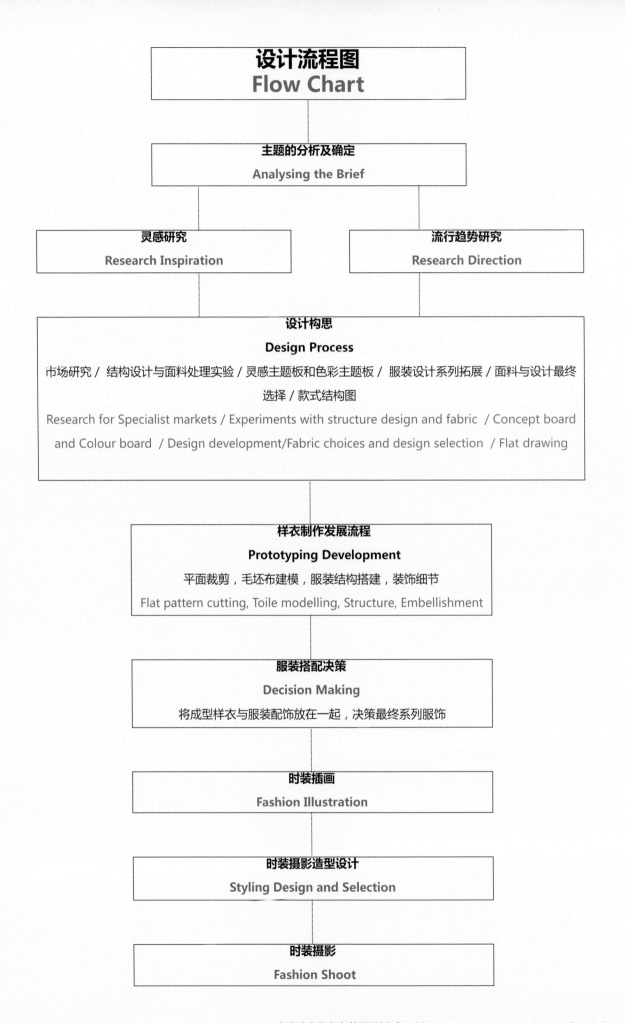

灵感研究
Research

Inspiration

个人灵感来源及主题研究
Personal inspiration and concept

奥菲斯（模型2，第二版）
1956, 芭芭拉·赫普沃斯
Orpheus (Maquette 2, version II)
1956, Barbara Hepworth

Pelagos，1946，
芭芭拉·赫普沃斯
Pelagos 1946,
Barbara Hepworth

波纹，1943-1946，
芭芭拉·赫普沃斯
Wave 1943-1946,
Barbara Hepworth

木材及弦线"赋格"，1956，
芭芭拉·赫普沃斯
Wood and Strings (Fugue) 1956,
Barbara Hepworth

赫普沃斯抽象雕塑研究
Research into Barbara Hepworth

　　本人是从赫普沃斯的抽象雕塑并研究分析其艺术特征开始这次毕业设计之旅的。研究的展开从一些小的细节开始，例如，分析赫普沃斯的抽象艺术特征；以及研究赫普的雕塑是如何平衡结构与细节。这些细小却又重要的特征细节非常重要，往往正是这些微小的细部特征，引发设计师们无限的创意。

　　在研究沃斯女士的作品时，本人深深的被她的雕塑所吸引（见左边的图片），它使人们联想到宁静而优雅的停歇在湖畔的天鹅；或是双臂舒展，在幽静的练舞房内翩翩起舞的芭蕾舞女演员。它是如此的令人陶醉着迷，引人浮想联翩。

I started with researching Barbara Hepworth's abstract sculpture and analysing the characteristics of Hepworth's art, which includes the figurative art, and the balance of structure and details in Hepworth's sculpture. I was attracted by Hepworth's sculpture (see images left), which reminded me of an elegant swan or a ballet dancer, who was stretching out their bodies in quiet, minimalist and elegant. They were fascinating to me.

弯曲的廓形 I
芭芭拉·赫普沃斯
Curved FormI, Barbara Hepworth

弯曲的廓形 II, 1955 年
芭芭拉·赫普沃斯
Curved FormII, 1955, Barbara Hepworth

椭圆形的雕塑 2 号, 1946 年,
芭芭拉·赫普沃斯
Oval Sculpture (No. 2) 1943,
Barbara Hepworth

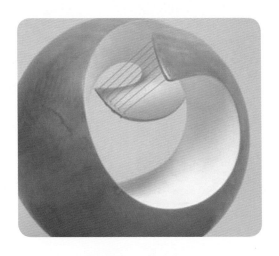

浮生物, 1946 年,
芭芭拉·赫普沃斯
Pelagos, 1946, Barbara Hepworth

从赫普沃斯（1943年，1946年，1956年）的作品中，可以发现流动优雅是她造型艺术的一大特征，这种独特的造型使得她的雕塑散发着典雅迷人的韵味，从而触发本人无限的灵感和遐想。

比如：从赫普沃斯的雕塑作品中（见左页），本人发现将雕塑实体造型做得光滑圆顺，结合自然或抽象的凹凸曲线，塑造出的雕塑作品，是极具有抽象艺术独特魅力的。她的雕塑通常由简单的几何形状组成，具有高度抛光表面和流畅优雅的曲线。她坚信将雕塑的实体部分设计成优美的曲线或造型是十分必要，因为这将会使观众对雕塑的外观留下深刻的不可磨灭的印象（赫普沃斯，1956年）。

From Hepworth's (1943, 1946, 1956) work, I found the figurative art is a strong character of her work, which made her sculpture very elegant, and inspired me to employ figurative art in the design of my garment at a later stage.

In her sculpture, I also found she liked to highlight the solid shape which she made very smooth, and combine the abstract and/or naturalistic characteristics in the convex or concave of curves, which is another characteristic of Hepworth's sculpture. Hepworth's sculpture usually has geometrically simple shape, which has highly polished surface with elegant curves and a variety of holes of different sizes. She believes that it is necessary to develop a sense of beauty for solid shape, or "form", because it will improve the appearance of the sculpture and impress the audience (Hepworth, 1956).

木材及弦线"赋格"，
1956，
芭芭拉·赫普沃斯

Wood and Strings
(Fugue) 1956,
Barbara Hepworth

其次，更值得一提的是在对沃斯女士的雕塑进行进一步的研究时，本人还发现她喜欢运用不同尺寸的孔洞来贯穿整体雕塑，形成一种特别但美观的雕塑结构，使得空气，空间和光线得以在这些孔洞里交相辉映，形成雕塑实体内的第二个虚拟的空间。这样的结构设计，空间的贯穿给予了传统雕塑新的生命和崭新的活力。

基于孔洞的设计，她雕塑的另一个显著的造型手法是对线条的运用。线条的穿透力和连接性可以很容易的使雕塑产生灵动的空间感，并且能够平衡实与虚之间的空间关系。例如，雕塑作品"木材与弦线的'赋格'"就将空间的力量与平衡处理得非常完美。雕塑品"木材与弦线的'赋格'" 是由一系列的抽象几何形体组成：强有力的线条和优美圆滑的孔洞巧妙的组合在一起，共同谱写出了由木头和弦线组成的节奏感很强的乐章。从这个雕塑中，人们可以看到有不同的设计线条从雕塑造型的一边紧密的联系着边缘光滑呈优美弧线的另外一边。"线"在这个雕塑作品中就像是一座桥梁，将两个不相关联，遥遥相对的两侧紧密的联系在了一起。而且，这个线不是普通的连接，它是经过周密设计的。沃斯女士（1962年）将其设计成了美丽的三维立体的网状，从不同的角度去看这个网状的"线条"的布局，欣赏者会随着探索的深入，观察手法的不同发现更多的趣味来。真是步步是设计，点点是细节。

What is more, she also liked using holes with different sizes in her sculpture to enlighten the structure for air, space and light to get through. Those holes gave breathing to her sculpture (Director, 1982).

Besides using holes, another strong characteristic of her sculpture is the use of lines. The lines are very important for making sense of space and making the balance of solid shape and the area that is not solid. For example, by observing the sculpture Wood and Strings 'Fugue', I perceived a combination of balance, strength and serenity of this sculpture that I was appealing to me. The sculpture 'Fugue' is composed of a series of shape: strong lines and holes which combine together and then compose as a completed one piece of art. From this sculpture, people can see different designed lines from one side of the sculpture, and smooth and curving edges of the wood from the other side of the sculpture. The lines in 'Fugue' look like a bridge connects one side to another. In addition, Hepworth (1962) has also made the lines as beautiful as a spider web.

线与造型的平衡

奥菲斯（模型 2，第二版）1956 年，赫普沃斯

The balance of line and form

Orpheus (Maquette 2, version II) 1956, Barbara Hepworth

孔洞与廓形的平衡
椭圆形的雕塑，1943，赫普沃斯
The Balance of Hole and Form
oval sculpture,1943. Barbara Hepworth

正是这样的散发独特魅力的造型以及巧妙设计的细节深深吸引着本人的眼光，使人产生无限的遐想空间。并启发了强烈的创作欲望。特别是通过"线条"与"孔洞"来平衡空间虚实的关系的触动，使本人产生了将其运用到时装设计中去的创意。对此本人展开了一系列的创作构思，草图勾画和艺术的联想。

　　例如，本人联想到"线"在本次时装设计中，其作用可以不仅仅是联系左右两块不同衣片，还可以通过线与线之间的交错变幻，形成一种随人体运动而变幻的设计面。"孔洞'也可以不仅仅是装饰的作用，人们通过它还可以欣赏到"第三空间"——也就是建立在人体自然曲线与服装廓形之间的空间美。设计师还将运用几何造型与不规则造型交织的强烈视觉感来设计这组成衣的具体廓形。这也就是本次毕业设计的主旨思想——创新与平衡——将赫普沃斯独特的艺术抽象美感导入服装设计当中，使时装也散发出雕塑艺术的魅力。

Hepworth's sculpture impressed me and inspired me for my own fashion design. I was impressed by the use of lines and holes to make the sense of art, and also employed the idea of using lines and holes in the design of my garment. In the application of the lines and holes in my fashion design, the lines can not only connect two blocks together, but also leave little space for people to see the shape of human body. The holes can be used for people to see the 'third space' which is the space between the garment and human body in my design. I will also use strong silhouette geometrically and asymmetrically in the parts of garments. Meanwhile, I will use curving silhouette, lines and holes to create and balance the shape of outside abstract garments with the outline of inside human naturalistic figure, which is the main concept employed in my fashion design.

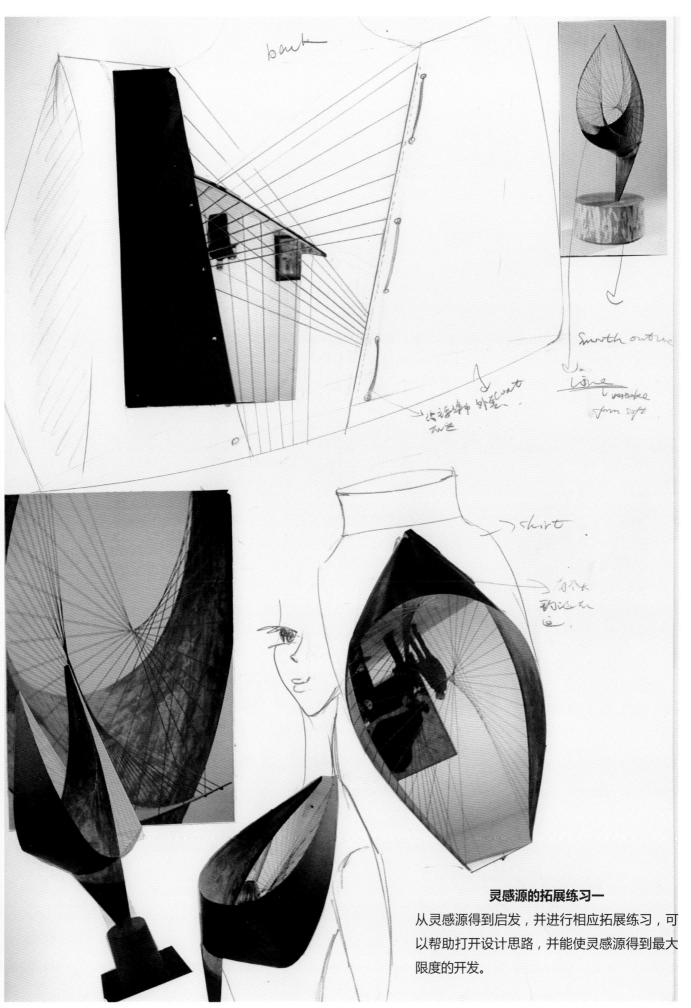

灵感源的拓展练习一

从灵感源得到启发,并进行相应拓展练习,可以帮助打开设计思路,并能使灵感源得到最大限度的开发。

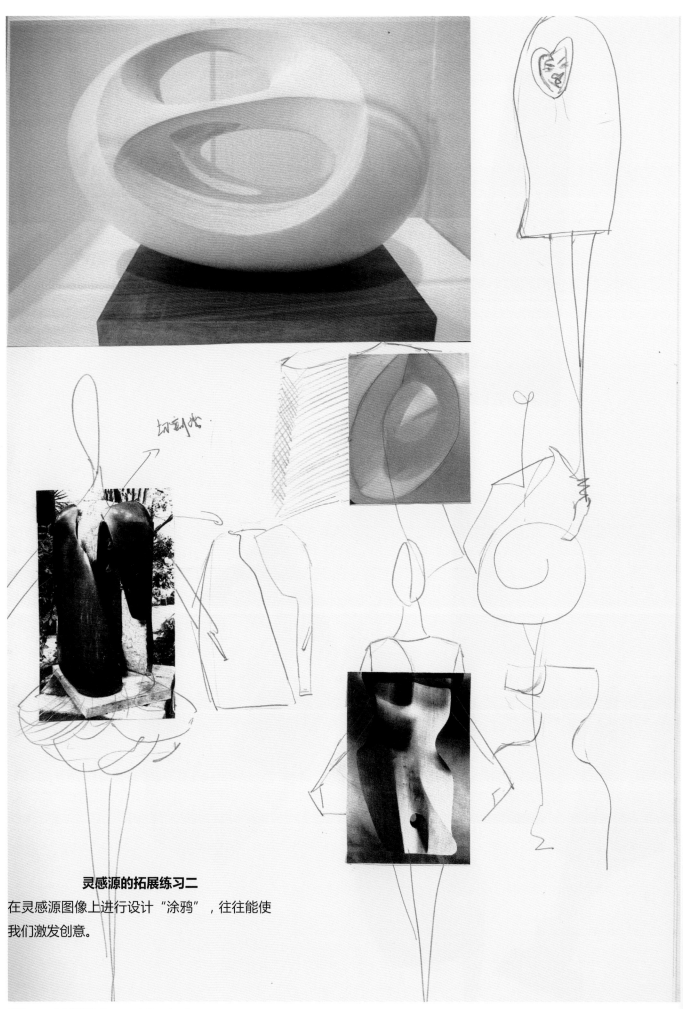

灵感源的拓展练习二

在灵感源图像上进行设计"涂鸦",往往能使我们激发创意。

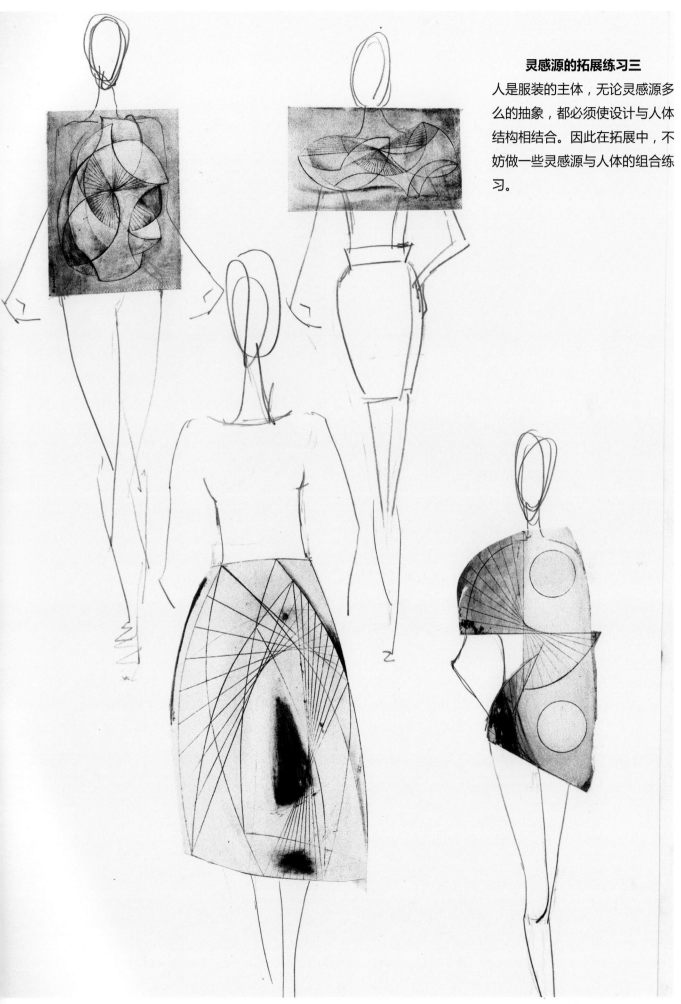

灵感源的拓展练习三

人是服装的主体,无论灵感源多么的抽象,都必须使设计与人体结构相结合。因此在拓展中,不妨做一些灵感源与人体的组合练习。

相同领域服装设计师们的研究与启发
Fashion Designers' Impact on My Design

通过对许多当代的著名时装设计大师的研究，例如：川久保玲，三宅一生，三本耀司和候塞因·卡拉扬。他们也给予了本次设计思路许多新的启发。

川久保玲是当今世界最有影响力的日本设计师之一。她改变了20世纪末21世纪初人们的穿衣哲学。她创造出了新的创意时尚——超大尺寸，信封式的廓形。她改变的不仅仅是传统的女性时装廓形，她还引导当代女性进行'无彩色，无性别'的着装革命。她的设计专注于研究人体与服装的关系，比如在女性的肩部和臀部做夸张放大的设计，这与传统的优雅高贵的法国高级时装廓形相去甚远。（清水，2005年）但是，也正是这样的大胆创新，使她的作品引导人们对人体本身曲线的思考。

本人对她的 Comme des Garcons 品牌在1997年推出的春夏时装"身体偶遇裙子，裙子邂逅身体"主题系列非常的感兴趣，由此展开了一系列的分析。其实，但看这个系列的名字就能让人浮想联翩，会使欣赏者不禁对一件裙子如何的与人体邂逅及亲密接触的过程产生一系列的联系。这组系列用紧身弹力材质与可拆卸衬垫的塑造了一系列独特有趣的时装造型。在这一系列中，川久保玲将这些紧身弹力材质包裹可拆卸衬垫设计放入服装的肩部、背部，身侧和臀部，从而缔造出了一种人们从未见过，夸张到极致的但又无比惊艳的廓形剪影。正是这组川久保玲的设计，使本人也产生了设计"当身体邂逅时装"的创作灵感。

The fashion design was also influenced by others designers, such as Rei Kawakubo, Issey Miyake, Yohji Yomamoto and Hussein Chalayan.

川久保玲　Rei Kawakubo

Rei Kawakubo is one of the world's most influential fashion designers. She changed the mind of people who lived in the 20th century. She found the new way to create fashion by designing oversized garment that envelope the female form rather than traditional, exposing and non-colour garment. Her design liked to investigate the relationship of human body with garment, such as exaggerating the part of shoulders and hips which was far from the heavy traditional French high fashion (Shimizu, 2005).

I am really interested in her Comme des Garcons' Spring / summer 1997 collection; the name of this collection is "Body meets Dress, Dress meets Body" The collection name is powerful and makes me think about the nature of physical beauty and femininity. Her collection includes a range of figure-hugging stretch garment, with removable down pads sewn inside the garment. Rei Kawakubo's pads were inserted at the upper back, side and shoulders, which creates extreme silhouettes, the likes of which had never been seen on catwalk (Fukai, et al, 2010a). Re Kawkubo's designs inspired me to research and create the relationship of "Body with dress" in my own design.

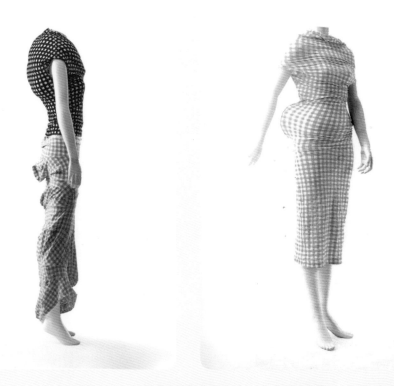

Comme des Garcons 品牌 1997 年春夏时装系列主题 "身体偶遇裙子，裙子邂逅身体"
Samples of "Body meets Dress, Dress meet Body" Comme des Garcons' Spring / summer 1997 collection

灵感研究 /Research Inspiration　031

三宅一生　Issey Miyake

在完成了对川久保玲的研究后，接着，本人对那些在处理服装结构与细节的平衡关系上的设计大师进行深入研究，比如三宅一生。

三宅一生最著名的是对可塑性褶裥织物的研发运用。这个褶裥织物是建立在人体结构上，并可以塑造出不规则的几何轮廓。三宅一生的时装设计与传统方式完全不同。他将剪裁的焦点放在了如何建立服装的结构和造型的比例上，以刻画一种"禅意"的织物雕塑感，被誉为最具有神圣东方哲学韵味的时装设计。除此之外，他还关注服装造型与细节的平衡（三宅一生，2010）。这在三宅一生早期作品中，尤其显著。他所塑造的一系列抽象几何褶皱服装，例如1994年春/夏季系列的"飞碟礼服"，就是用几何剪裁切割与皱褶的特性，去塑造一个新的、完全打破传统规则的廓形，而且这个廓形会随着人体的运动方式产生不一样的变化。甚至于，人们在不同的角度欣赏同一件衣服，它所产生的廓形美感也会不同。（布莱克，2006a，2006b）

Besides Rei Kawkubo, I take a close looking at the designers who focused on the balance of structure and details, such as Issey Miyake.

I was inspired by Issey Miyake who was famous for using pleating fabric to form geometric outline, which was built in human structure. Issey Miyake's design ignored traditional cutting, but focused on the proportions between clothes and figurative sculpture. In addition, he was also successful in balancing garment figure and fabric detail (Miyake, 2010). For example, in Issey Miyake's early collection, he created a series of dynamic garment by using abstract geometric form with pleating fabric, such as "Flying Saucer dress" in Spring / Summer 1994. The garment used geometric cutting and the characteristic of pleats to create an abstract form, in order to create different shapes whilst movement and be appreciated at different points of view (Black, 2006a; 2006b).

三宅一生在 1994 年春 / 夏季时装周设计的 "飞碟礼服"
The design of "Flying Saucer dress" in Spring / Summer 1994, Issey Miyake

另一组带来强烈灵感冲动的三宅一生的作品是：2010年三宅一生的"1325"主题系列———这个系列以折纸手工和可持续性理念为基础。该系列名称中的数字1是指每件衣服由一块布料做成,3是指衣服呈三维立体形状,2是指衣服可以被折成二维形状,5是指每件衣服有多种穿法。(三宅一生,2010)它是通过人体的体型来撑出衣服的三维立体的廓形,本体的服装其实只是一片经过独特设计的面料。这个概念触动了本人很好的设计思考：如何运用面料所本固有的特性,设计出建立在人体结构学上的时尚服装？

Another example of Issey Miyake's creative design is 'A piece of flat material become a three-dimensional structure' in his "1325"collection, 2010. The structure in turn can become a two-dimensional shape with the addition of straight lines. But when human wear this piece of flat material, it can be a new different shape based on the human body shape (Frankel, 1997; Miyake, 2010). This also inspired me in my own garment design in terms of how the garment can build in the human body structure while taking maximum advantage of the fabric's own feature.

2010年三宅一生的"1325"主题系列
A piece of flat material become a three- dimensional structure'"1325"collection, 2010, Issey Miyake

山本耀司　Yohji yamamoto

　　与三宅一生设计理念相似的日本设计师是山本耀司，他被誉为日本时尚界的教父，他同样在时装设计中注重服装结构与细节的平衡关系。

　　他的服装有其独特的抽象感，特别是他服装设计中的对称感与不对称剪裁的平衡关系非常的令人着迷。这种特殊的对称性在他的时装中也不是传统意义上的对称感，而是用不对称的裁剪设计出一种人类视觉上的对称平衡。例如：在2001-2002阿迪达斯／山本耀司秋冬发布会中，他所设计的经典Y-3条纹海军蓝羊毛华达呢拉链外套就充分体现出了他用不平衡来平衡视觉的设计思路（见后页）。衣服右侧设计成由经典Y-3条纹装饰的完全束缚不可动的大袖子，而另一侧则设计成深蓝色简单的可灵活活动的紧到极致的左袖。他用不平衡的设计——用紧而灵活的左袖去平衡松而束缚的右袖，又用大面积但深沉的蓝色去碰撞小面积但活力四射的条纹——在视觉上反而塑造出一种惊人的和谐感、平衡感。

　　除了用不对称的剪裁设计来达到视觉上的平衡，"简约"是三本耀司时装设计的另一个显著的特点。

Similar to Issey Miyake, Yohji Yamamoto is another famous designer focuses on the balance of structure and details.

His clothes are in a scene of peculiarly abstract. The way how he balances the asymmetry and symmetry in his design fascinated me. Symmetry in his garment is not in a traditional definition; in fact, the symmetry is reflected by his application of asymmetry to make a visual balance of the garment. For example, in Adidas /Yohji Yamamoto Autumn/Winter 2001-02, he designed the Navy-blue Wool Zip-opening Gabardine Jacket which has a white striped inlay in the right sleeve which is restricted with the body jacket. Other parts of the garment are totally in blue colour with a tight left sleeve. The balance of symmetry and asymmetry reflects on the colour and sleeves. In terms of the colour, the large area of blue balances the small area of white strips. With regard to the two sleeves, the tight left sleeve balances the overall constrained feeling of the garment caused by the restricted right sleeve (Fukai et al, 2010b).

Besides the feature of a balance of symmetry and asymmetry, simplicity is another feature of Yamamoto's design.

海军蓝羊毛华达呢拉链外套

The Navy- blue wool zip-opening gabardine jacket

他的时装都有一种独特的设计美学，叫做"残缺美"。因此，你在他的作品中会发现，他摒弃了一切其他设计师增添的繁复细节，使他的作品散发出"简约"的魅力。此外，他服装的简约雕塑感是完全建立在人体工程学上的，这点可以从 1996/1997 年山本耀司秋冬秀中的白色毡步连衣裙（见后页）中得到反映。这件裙子，他重点突出了礼服的背面。他用精湛的剪裁和精致的细节处理服装背部的结构，从而凸显了裙子前片的简约感。而裙子后背的开口，结构线的弧度处理，面料材质的运用，面料如何在人体上站立的方式等等，都是建立在设计对人体工程学的理解上。使看似简单的时装作品散发出雕塑才具有的艺术气息。以上这些也正是我从对山本耀司的作品中所感悟的，即建立在人体结构上的简约，和用不对称的剪裁设计来达到视觉上的平衡。

His design employs the concept of 'aesthetics of the incomplete', so he cut off all the unnecessary details rather than adding details as some other designers do. Additionally, his sculptural form is based on human being's natural structure, which is reflected on the Yohji Yamamot Autumn/Winter 1996-97 White Wool Felt Dress .Of that dress, he emphasises more on the back of the dress, where he used cutting and detailing to balance the simple front of the dress (Washida, 2002). Therefore, Yamamot 's design taught me how to create a balance and how to keep simplicity in my own fashion design.

1996/1997 年山本耀司秋冬秀白色毛毡连衣裙
Yohji Yamamot Autumn/Winter 1996-97 White Wool Felt Dress

在服装材料的研究上，本人也调查研究分析了一些善于利用材料特性或善于运用特殊方式使用服装面料的服装设计师，比如像是候塞因·卡拉扬．

从候塞因·卡拉扬历届的时装秀中的作品，可以发现他在服装设计中，非纺织类材料运用之多，材料的选择范围之广，创意的点子之奇，是一般设计师难以比拟的。例如，在他的2000年春夏时装发布会上的"飞机"礼服（见后页）的材质就是由玻璃纤维压膜制成的，玻璃纤维立体成型形成了光洁而简约的廓形，变成一个雕塑般独立的"身体"。在这光洁的玻璃纤维表面下，是有网眼纱层层叠叠组成的蓬裙。当模特展示时，雕塑般的表面在高科技的控制下慢慢升起，里面的网纱蓬裙便柔美的演绎浪漫的风情（奎因，2003：67）。从候塞因·卡拉扬的设计中，又引发了本人许多灵感，设想在毕业作品系列中运用一些意想不到的的材质来设计雕塑感般的造型。但是如果选择这样的方式，那么，如何去缝制这种非常规材质就会成为一个新的挑战。对此，本人更加深入的对候赛因的缝制的处理方式，进行了相应的研究。研究结果表明，由于候赛因几乎是运用硬质材质的材料进行设计，他一般是将两种材质进行不同的处理。纺织类面料还是遵循常规的缝制手法；非常规面料则会动用焊接，压膜等高科技手段。这一系列的缝纫研究给予本人许多启发，那就是将不同材质，分别用不同的拼贴手法处理。不过，其中非常规类材质的处理方式，并不符合整体的时装设计宗旨。因为本人的设计目标是设计制作出一组高级成衣系列，那么过于非常规的处理方式可能在成衣系列中并不是那么的符合成衣消费市场。因此，非常规类材质如何与普通面料进行缝合，会在之后的实验中成为最为重要的实验项目。

候塞因·卡拉扬
Hussein Chalayan

In terms of the material, there are designers who combine traditional and modern method to make good use of the special materials. The typical one is Hussein Chalayan.

In a collection of Hussein Chalayan's fashion work, I found that he used a broad range of unusual materials to design garments. For instance, in his Spring/Summer 2000 show, "the 'aeroplane' dress was made of moulded fibreglass, with moving panels revealing a froth of tulle. The fibreglass panels of this dress form a dense structure that maintains its shape independently of the body." (Quinn, 2003: 67). Inspired from his design, I thought my collection could use some unexpected materials to make the shape of garments, but how to sew these materials could be a challenge. Chalayan usually screwed the wood and other hard materials together, which would not be applicable to my fashion design because my aim is to create a ready-to-wear collection. Therefore, the choice of materials needs to at the convenience of sewing methods.

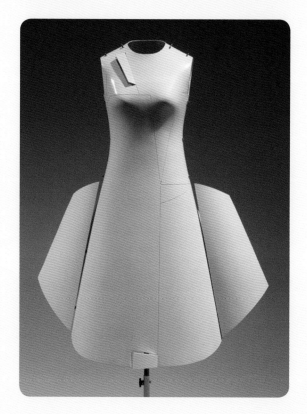

2000 年春 / 夏系列 ' 飞机 ' 礼服
The 'aeroplane' dresses spring/summer 2000

灵感研究 /Research Inspiration

基于对上述艺术家及相关设计师作品的调查和研究，接下来，本人针对高级成衣目标客户群做了市场调查及流行趋势调查。本次高级成衣的目标客户群的年龄层是28-38岁的时尚成功女性。根据ERAY高级营销公司2009年-2010年的市场调查报告中分析，针对这类女性的服装色彩，面料需求，款式喜好，主要社交场所等特征进行了系统的分析。数据表明28-38的时尚成功女性，主要社交场所是一些正式的中高端场所，比如像是展览或鸡尾酒会。着装颜色偏向于细腻和淡雅的色彩，如白色，浅灰色，裸体和黑色。面料选择则倾向于舒适，自然和高品质的面料，如100%桑蚕丝，100%纯棉弹力面料。设计款式风格则多为简约、优雅的时尚设计。

Based on the ideas of other practitioners as discussed above, the target customers of my garment were accordingly targeted. Also referring to the marketing report of ERAY Company 2009-2010, my target customers were focused on the following attributes: age, venue, colour, fabric and style. The age of the target customers would be between 28 and 38 mature females; the occasions when they wear my garment would be for formal and smart functions, such as exhibition and cocktail party. Colour of the design needed to be delicate and subtle colour, such as white, light grey, nude and black to show its elegance. Comfortable, natural and high quality fabric, such as 100% silk , 100% cotton and stretch fabrics would reflect the elegance. The style needs to be simple but considered design.

流行趋势研究
Fashion Trend Research

注意观察时尚媒体杂志或网站,并分析、概括时尚资讯是必不可少的灵感研究环节,它可以帮助我们解读服装流行趋势,以便在设计中能将灵感源与最流行的元素进行最佳的结合。使设计与工艺不会偏离市场。

2011年欧美春夏发布会,设计手工细节充满了摩登的色彩(见上图),而且这是2010年秋冬季设计编织细节的延续。设计师运用现代摩登的设计将古老的手工工艺技术穿插其中,如:打结,编织,镂空和刺绣。不仅构建了外观精美的手工制作服饰,更是重现了的历史手工特殊工艺,并使服装更显精美。

从英美两大时尚资讯权威网站 vogue.co.uk 和 style.com 得到的时尚信息来看（见上图）。设计师用纯白、象牙白、裸色，以及各种淡雅的颜色编织了 2011 春夏的浅色世界。浅淡的色彩彰显着低调的奢华和别致的优雅。以白色的纯净来沉淀心情；以裸色来挑高身姿。或许是受到 2012 世界末日的影响，时装中也透视着人性心理的因素。清爽的色彩更是将一些微小的细节展露眼前，诱惑着人们去慢慢品味。

灵感研究/Research Inspiration

设计构思
Design process

结构设计与面料处理实验
Experiments with Structure Design and Fabric

本次的设计是从对服装面料的选择开始的。在经过层层实验和筛选，本人择取了白色的重型无纺织面料来塑造整体的服装造型。重型无纺织面料不是常规的服装制作面料，它通常被用于制作工艺品，帽子坯样，或当粘合板给灯具做装饰。不过，本人选择它作为服装主打面料，是因为它能够提供足够的塑造性和韧性。但是，在前期的塑形实验中，本人发现，这种面料虽然极容易塑形，轻易可以达到设计师所追求的雕塑感的廓形；但是，由于它并不属于常规性纺织类材料，材料硬度过大，并不适合贴身穿着，而且，它容易造成人体活动的不便，活动受限制。当本人尝试着将其裁剪成普通服装的结构，并进行着装活动实验，着装者的手臂在袖筒里不能自由弯曲，活动极其不便。

为了解决这问题，本人进行了更多的面料重塑实验，希望这类面料在满足我塑形要求的同时，又能满足人体活动的基本需求。首先，设计师对这种重型无纺织布进行褶皱处理，希望能将其变得柔软。但是结果并不理想。普通柔软面料在进行褶皱处理时所能达到的清晰的褶皱感，在这个面料上得不到呈现；反而这种模糊的褶皱感破坏了材料本身的塑形优势，使造型变得凹凸不平，不具备赫普沃斯女士雕塑作品的圆润感。这不是本人所期许的。

I started with choosing materials for my design. I chose white heavy non-woven fabric to create the shape of my garment. This fabric is usually used for crafts, pelmets and hats, as well as interlinings or foundation, because it can provide specific functions, such as resilience and strength. Therefore, using this fabric would create and maintain the silhouette of my garments. However, during my experiments, I realised that it was easy to make shapes with it, it was difficult to wear, because it was very inflexible. When I tried to wear it, my arms could not bend. In order to solve the problem and make it flexible while flexibility was also required to be stiff. I, firstly, pleated the material and made it softer; however, the result was still not suitable for my design, because the outline of my collection needed to be strong and clear. Pleating the fabric made the outline unstructured.

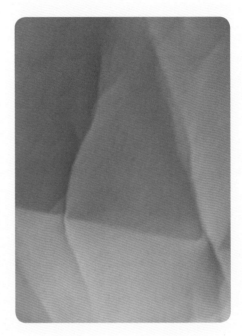

重型无纺织布的褶皱试验效果
Pleating the Hard fabric

柔软面料褶裥试验效果：
Pleating the soft fabric

褶皱实验失败后，本人马上针对所出现的问题，做了另一个面料重塑试验（见下页），试图解决目前所遇到的问题。在这次试验中，本人用一把美工刀将这种重型无纺织布切成一条条间距 0.3cm 宽，线条宽度也是 0.5cm 的细线，以此来消除面料的僵硬感。之所以采用 0.5cm 的宽窄的线条剪裁，是因为通过实验证明，0.25cm / 0.3cm 的宽窄会使面料的可塑性大大降低，廓形的韧性也会下降，并且面料的着装牢固度也会降低。而采用 0.7/0.8cm 宽窄的剪裁，牢固度得到了加强，但是线条太过粗放，精细度不够，不符合赫普沃斯抽象雕塑的优雅感。而 0.5cm 宽度比例，最能完美的满足廓形需求，同时还能够符合正常服装的牢固需求。

　　更为美妙的是，通过的精确的线条剪裁，每一根线条韧性不同，在人体穿着时，每一根线条会随着人体活动的变化而变化，使服装廓形有一种变幻动感的梦幻感。这种面料重塑设计选择是明智的，它与其他简单的二次面料改造不同，它是为人体，为服装的整体造型服务的，既是服装的廓形，又是服装的细节。人们甚至可以通过线与线之间的空隙欣赏人体曲线独特的韵律美感。因此，这种面料的重塑改造工艺对于本次的成衣系列设计而言，无疑是成功的。

　　在面料调试试验后，本人调整了的最初的服装设计草图。我决定将线条作为我服装的主要细节元素。当设计思路变化后，相应的，本人急需要解决的是：如何将这样的外观切割线融入到服装结构设计当中去，使服装结构与细节得以平衡。

After that, I did another experiment with the fabric. I used a knife to cut the fabric line by line. The space between the lines is only 0.5 centimetres, which has an allowance between one line to another, which can also help the movement of human body. In addition, every line can make different shape when wearer moves because of the fabric's function included in the collection. The audiences can even see the curve of human body through the allowance. The result of cutting lines in the fabric proved to be very successful.

After the moderation of fabric, I modified the draft design of the garments. I decided to use the line to become the main element of my garments' detail. Then, In order balance the solid structure with details. I needed to resolve how to integrate the cutting line into the overall look of the garments.

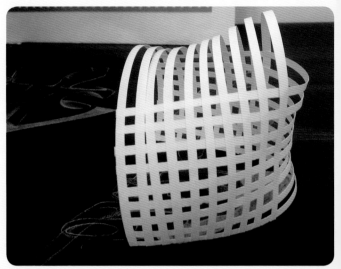
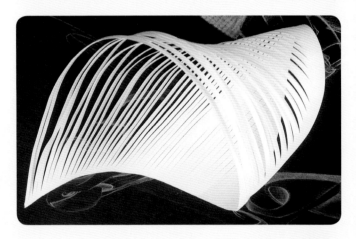

面料实验二：
线切割后的效果
The fabric experiment :
Cutting the line

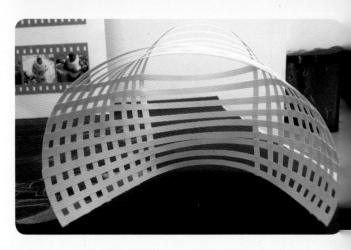

从三本耀司的服装作品研究中（奎因，2003年，布莱克，2006年）本人发现他的设计细节都是建立在人体结构分析的基础上的。因此，每一样细节设计在他的时装系列中，不但不会显得突兀，而且还使整体服装锦上添花，显示出低调的奢华。因此，我就用"犀牛"三维模拟软件，运用一些基本人体活动数据，构建人体模型，模拟人类正常走路时的状态。得以了解人体在走路时，人体内在结构的基本变化。

从这个模拟实验中（见下页）可以得知：人体上的每一个关节点是支撑人体正常走路的主要运动部位。这也就意味着，在服装设计当中，只要保持每个运动关节点的活动自如，也就满足了人类基本的活动需求。由此，本人将线条切割设计布局在肩点，肩胛点，腋点和肘点等人体关节点处。通过这样的设计布局，就可以解决穿着重型无纺织布无法进行人体基本活动的难点了。

From Yomamoto's garments (Quinn, 2003; Black, 2006a), I found his design detail is based on the human body structure analysis, so the details appear in his design is not abrupt but looks very elegantly luxurious. Thus, I used the software (Rhino) to simulate the movement of the human body, to understand the changes in the structure of movement.

From this simulation, I found the joint of human body is to help human movement. Thus, I arranged cutting lines in the areas of shoulder, back, armpit and elbow and used tension of lines in these area to help with movement. Moreover, I reviewed Hepworth's sculpture 'Wood and Strings (fugue)1956' (The Tate Gallery, 1982; Thistlewood et al, 1996), and felt the sense of femininity, smooth and soft line definition, such as curve, roundness and holes which are characteristics of her work. Accordingly, I designed these elements in details of my garments such as neckline, garment bottom, sleeve and wrist.

人体结构运动模拟的研究实验
The experiment of the the human body movement

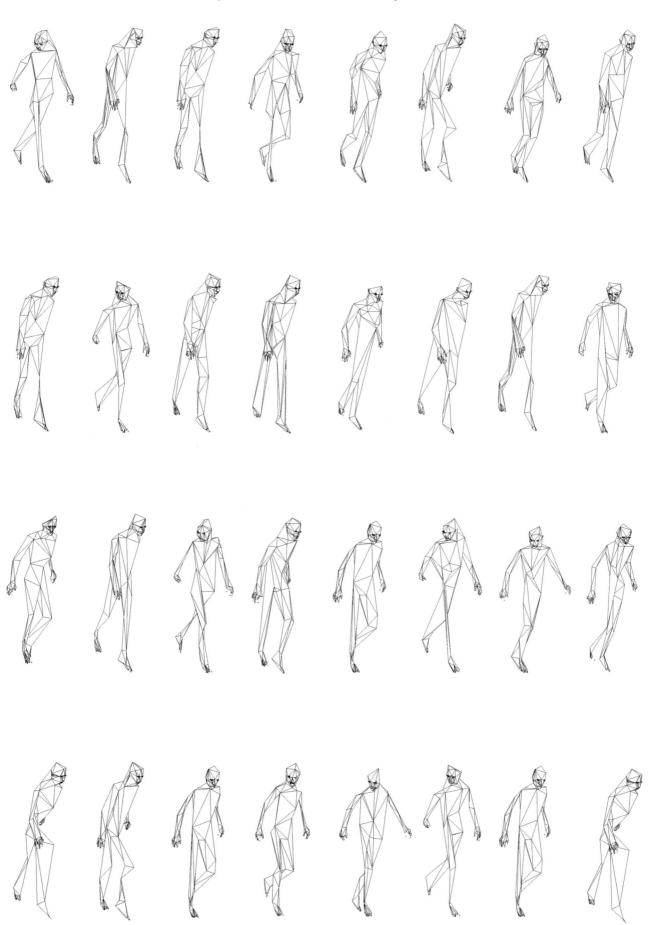

050 设计构思 / Design Process

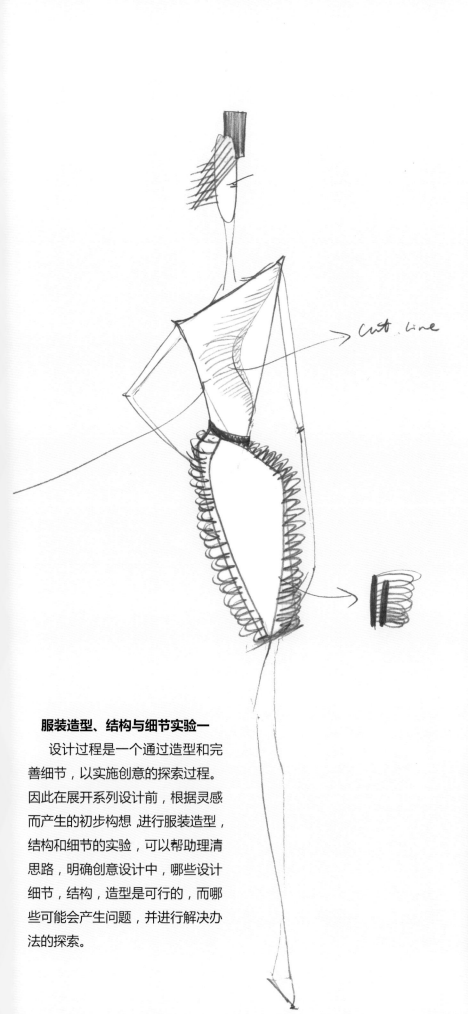
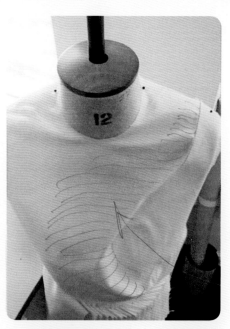
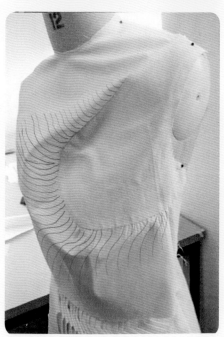
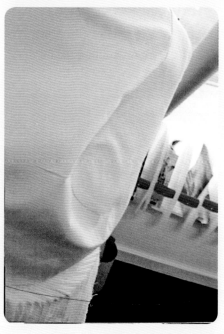

服装造型、结构与细节实验一

设计过程是一个通过造型和完善细节，以实施创意的探索过程。因此在展开系列设计前，根据灵感而产生的初步构想，进行服装造型，结构和细节的实验，可以帮助理清思路，明确创意设计中，哪些设计细节，结构，造型是可行的，而哪些可能会产生问题，并进行解决办法的探索。

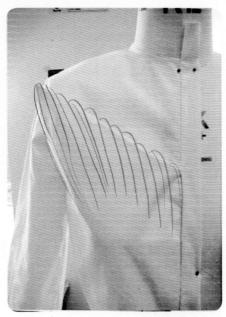
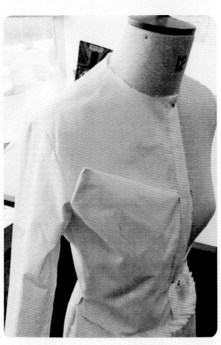

服装造型、结构与细节实验二

从不同的方面进行细节实验，如服装的立体感、服装的结构线设计、并将细节体现到造型上去等等。在实验二中，如何将服装呈现立体的空间感造型，以及将面料实验的结果作为细节在服装结构中体现出来。

有目的的进行细节实验，可以让设计者在今后的服装系列设计以及样衣制作中事半功倍。

服装造型、结构与细节实验三

这是几张探究廓型、款式和细节的实验草图。在坯样上进行细节构思，使创意能呈现出实际效果，并对服饰最后的形成在脑子中有了图像。

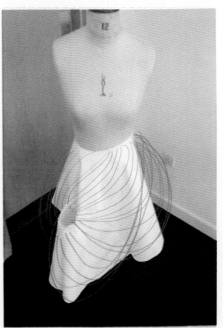

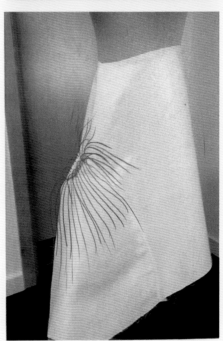

设计构思 / Design Process

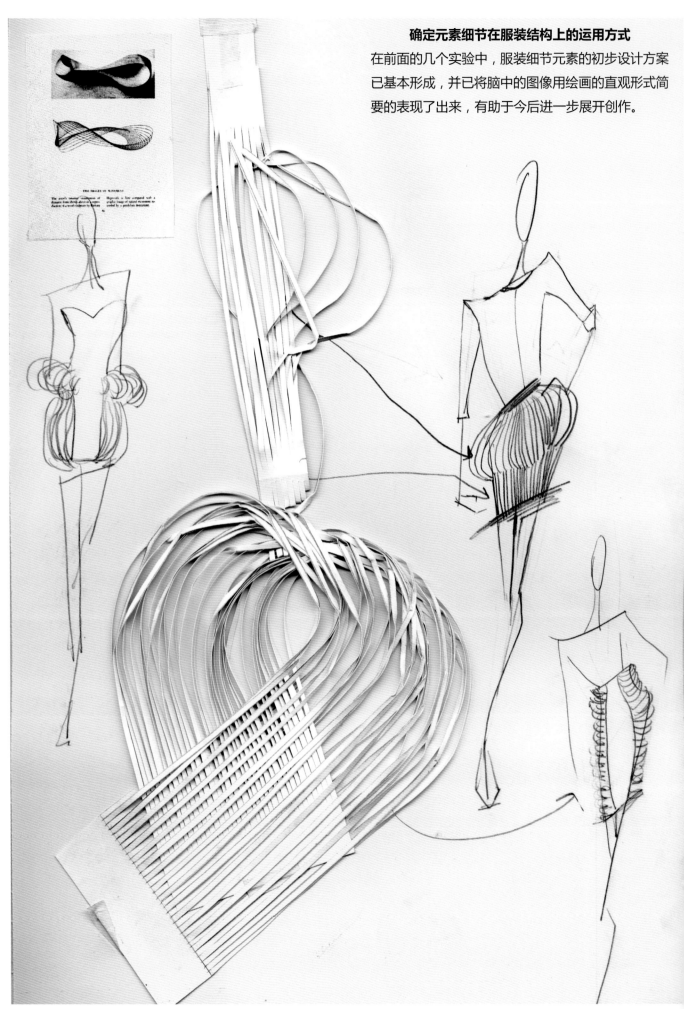

确定元素细节在服装结构上的运用方式

在前面的几个实验中,服装细节元素的初步设计方案已基本形成,并已将脑中的图像用绘画的直观形式简要的表现了出来,有助于今后进一步展开创作。

灵感和色彩主题板
Concept and Colour Board

　　灵感和色彩主题板可帮助设计师对选好的素材的进一步研究提供重点对象，并且帮助设计师有目的地对精心选出的目标主题进行设计。

　　这两块主题版，呈现了艺术的，雕塑的，优雅的但又现代的摩登感。主题板的设计有许多种方式，可以用传统的元素图片拼贴的办法，也可用一些设计软件达到所设想的效果，如：Adobe photoshop，Adobe Illustration 或 CorelDRAW。

Concept and Colour board can help designer to research inspiration deeply. It also can also help designer to know which the main point of the collection is, and how to pick up the real useful inspiration into garment design.
Both of concept boards show the sense of artistic, modern, sculptural and elegant.

色彩主题板

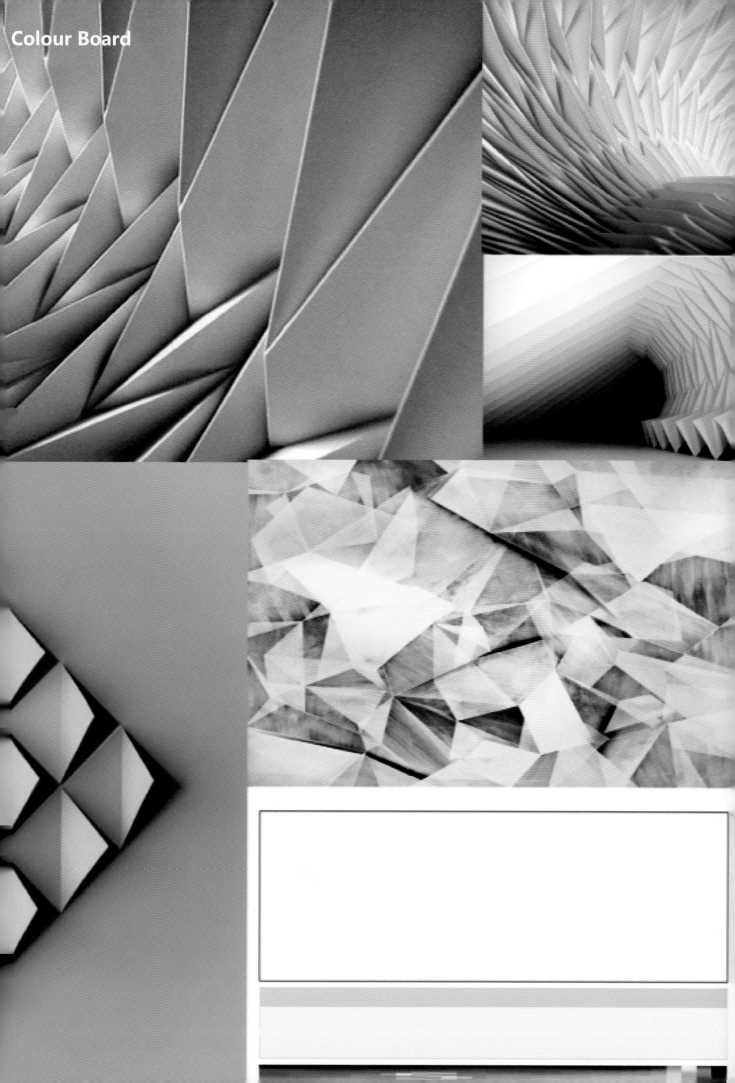

服装系列设计拓展
Design Development

  根据以上灵感源研究以及面料、造型、结构和细节的各种实验，本人拓展设计出了八套强调实与虚空间感，并具有简约优雅线条的高级成衣。（见下页的图片）。由于时间的限制，最后，从中挑选出四件作为本次研究生毕业设计的最终制作系列主线，进行更加深入的优化细节，坯样制作。

Bearing in mind of the inspiration from influencing practitioners, I have designed 8 sets of ready-to-wear garments which have strong simple outline and mixed solid forms and negative spaces (see image next pages). Due to the time limitation of designing the garment, 4 designs were chosen to form a collection for this project design.

系列设计拓展草图方案一

将系列构思草图放在一起,通过比较,可以整体的审视服装是否呈系列性,以及与灵感主题是否符合,这样就可以更有效的进行修改调整。

设计构思 / Design Process

系列设计拓展草图方案二

草图二中侧重于审视一些细节特征,看是否与灵感创意相衔接,并思考在制作上的可操作性。

最终系列设计拓展草图方案

设计构思 / Design Process

最终阵容
Line Up

就是定型的系列设计款图，最后阵容的确立必须使设想化为现实的过程中具有可操作性，并使款式、色彩、细节等配搭形成完美的创意概念

面料的选择与搭配
Fabric Choices and Selection

面料的选择及如何搭配应用也是重要一环。通过对伦敦面料市场的调查采集，结合所针对的消费层，最终选择了含 100% 丝的针/梭织面料做底衫；含羊毛的重型无纺织布做廓形构建。在选定面料的过程中，尤为重要的是面料搭配。按设计意图裁剪，并在纸模特上直接搭配调整，优选出符合灵感创意的面料组合。

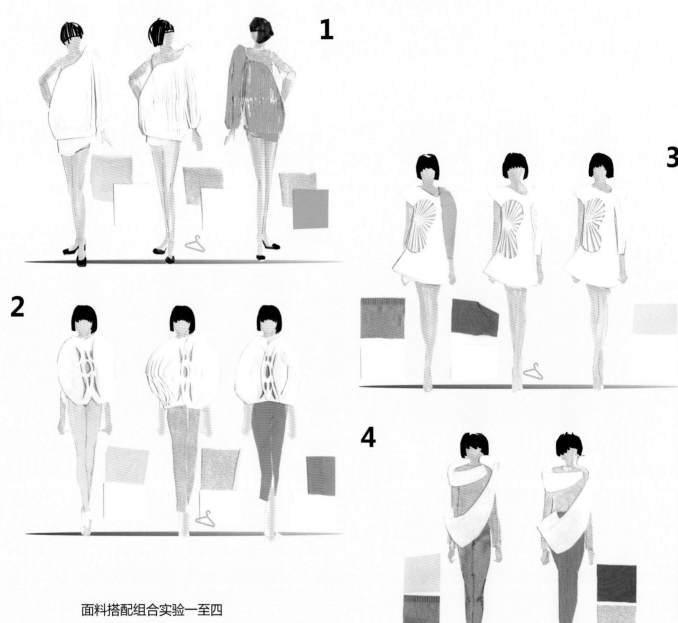

面料搭配组合实验一至四

利用纸模特进行面料搭配实验,什么样的搭配最符合原本设定的主题效果,变的一目了然。

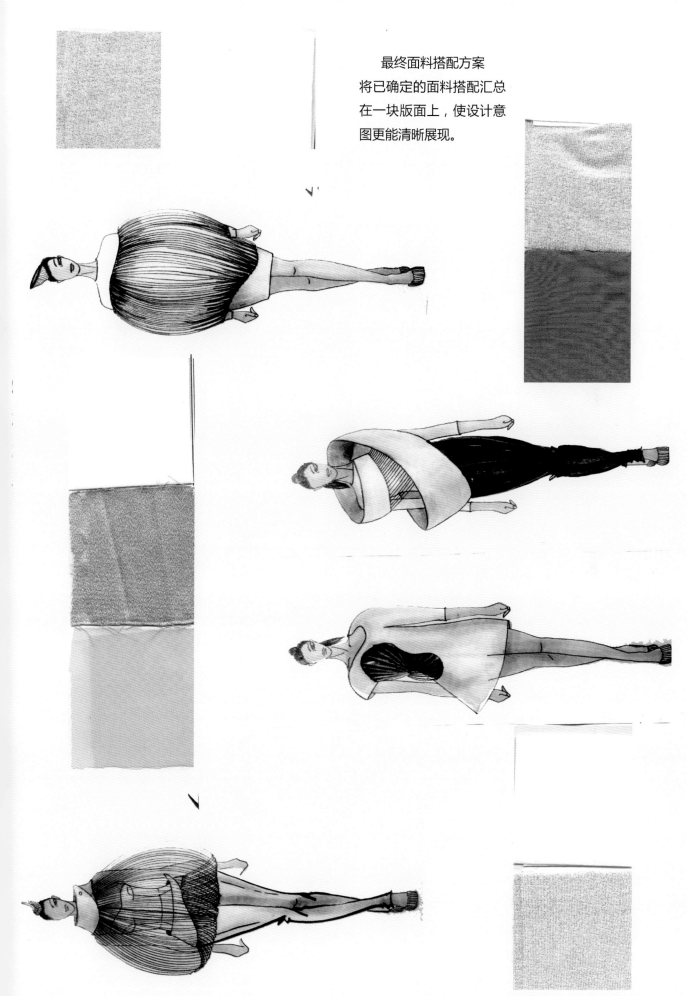

最终面料搭配方案
将已确定的面料搭配汇总在一块版面上,使设计意图更能清晰展现。

样衣制作及发展流程

平面裁剪

立体裁剪

服装结构构造处理

装饰细节处理

Flat Pattern Cutting

Toile Modelling

Structure

Embellishment

第一件样衣制作发展流程
The Development of The First Prototyping

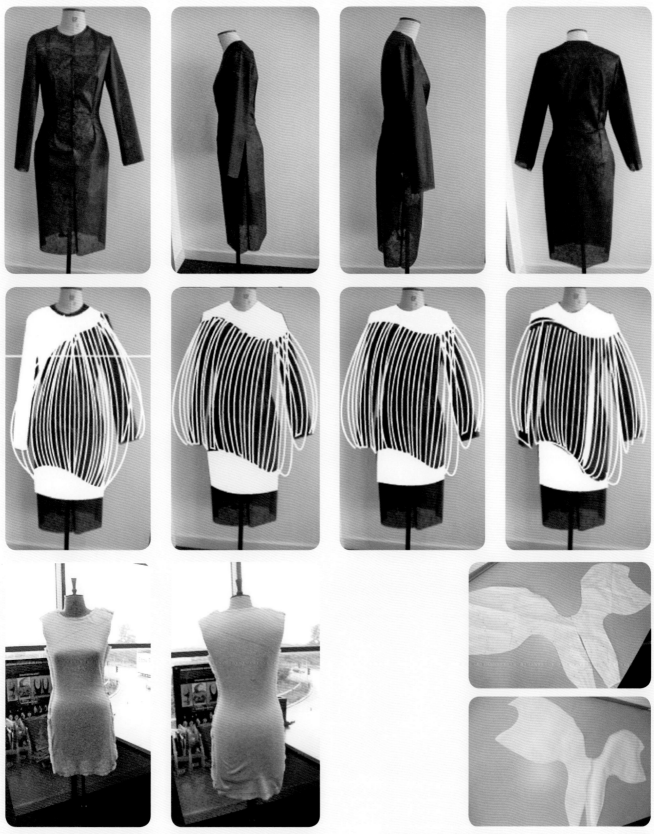

样衣制作及发展流程 / Prototyping development

070　样衣制作及发展流程 / Prototyping development

第二件样衣制作发展流程
The Development of The Second Prototyping

样衣制作及发展流程 / Prototyping development

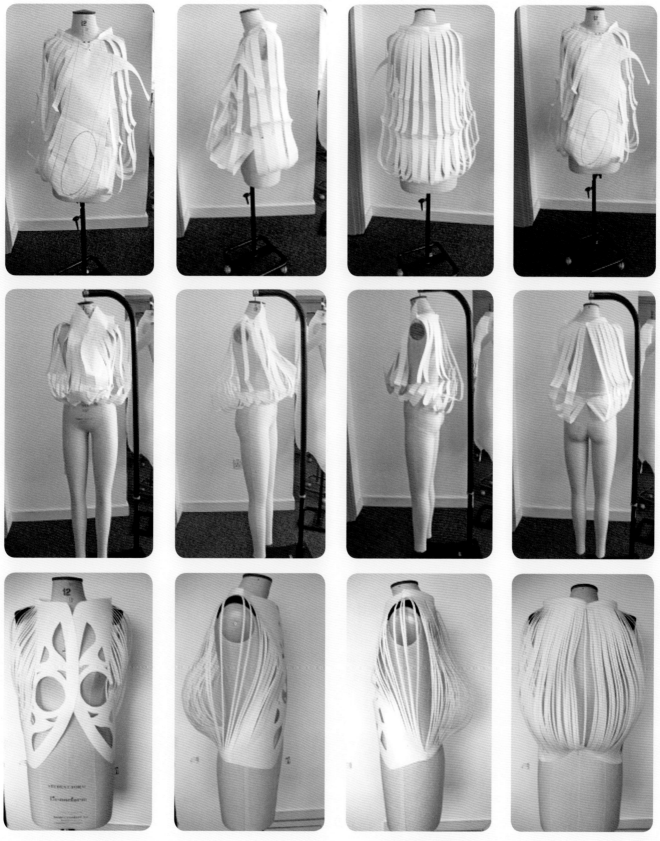

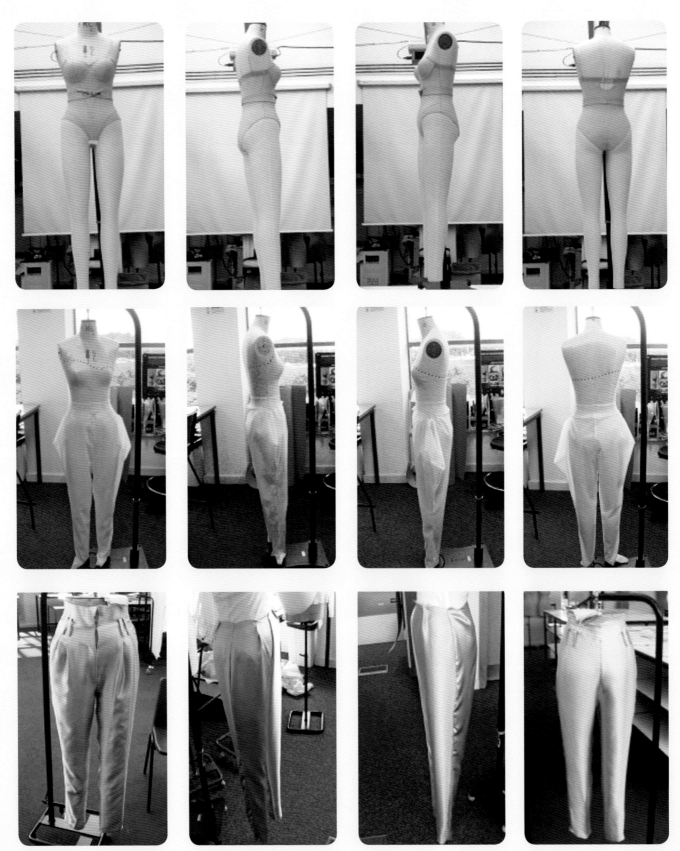

样衣制作及发展流程 / Prototyping development

第三件样衣制作发展流程
The Development of The Third Prototyping

样衣制作及发展流程 / Prototyping development

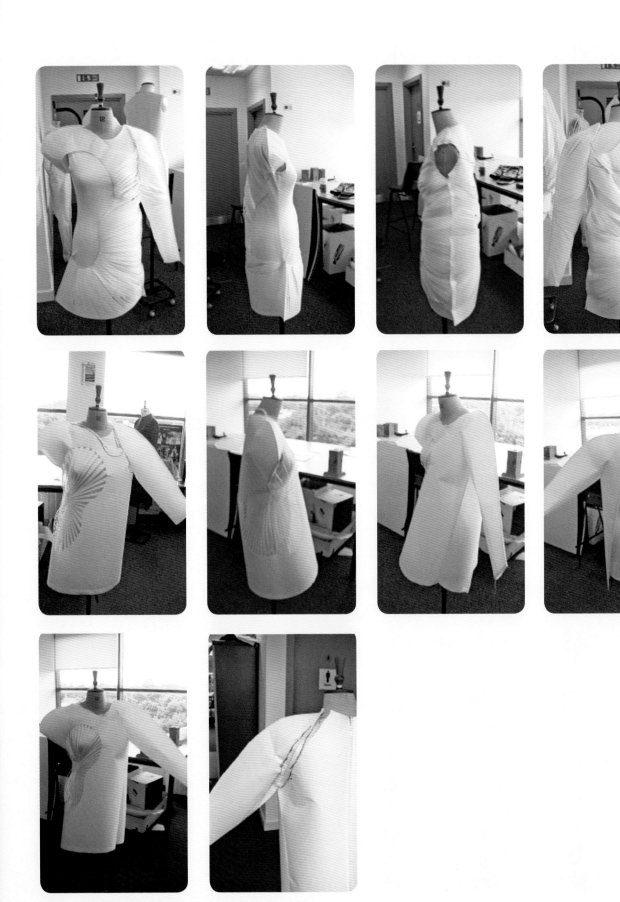

样衣制作及发展流程 / Prototyping development

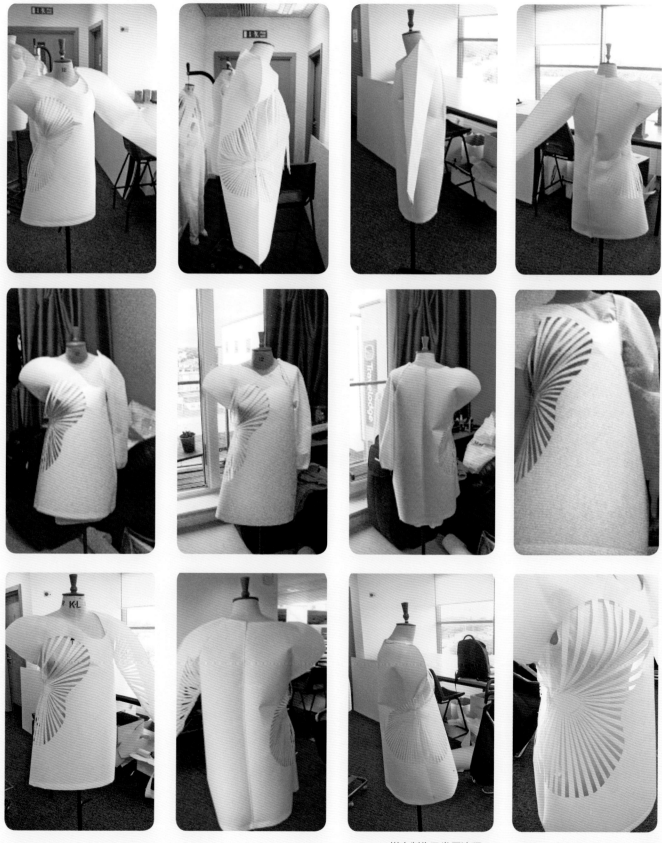

样衣制作及发展流程 / Prototyping development

第四件样衣制作发展流程
The Development of The Fourth Prototyping

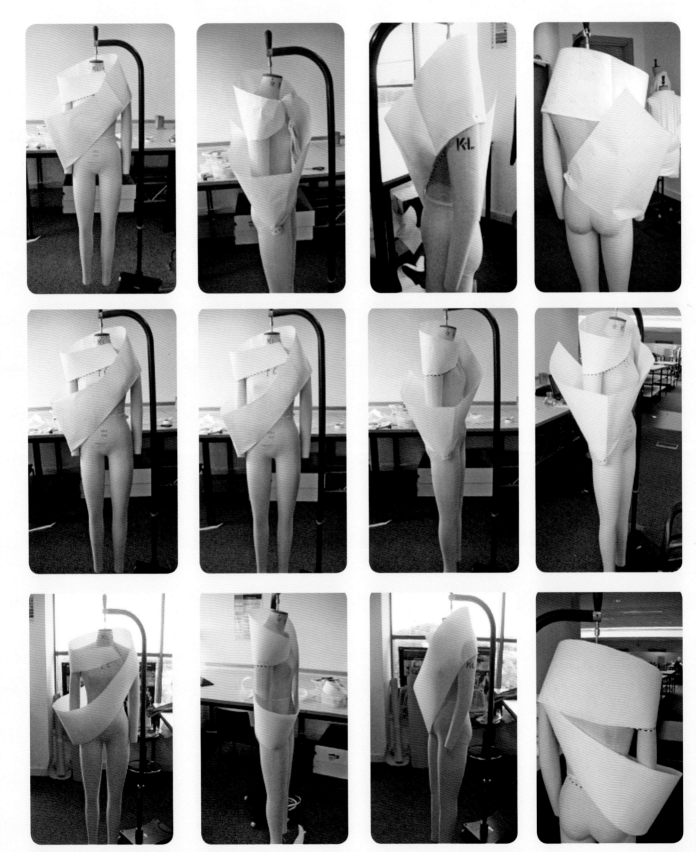

076　样衣制作及发展流程 / Prototyping development

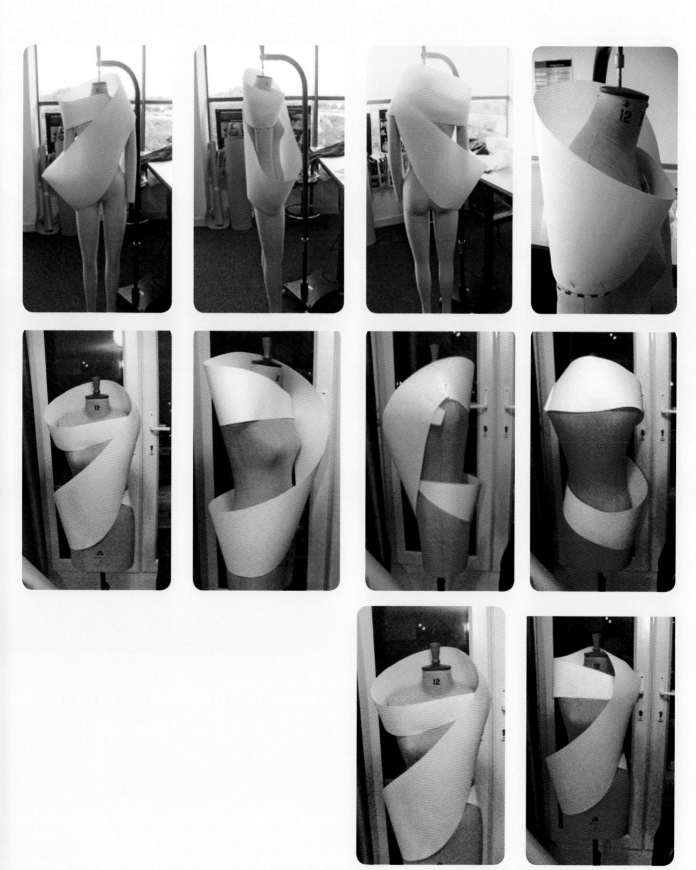

样衣制作及发展流程 / Prototyping development

组合样衣审样
The Group of Prototyping

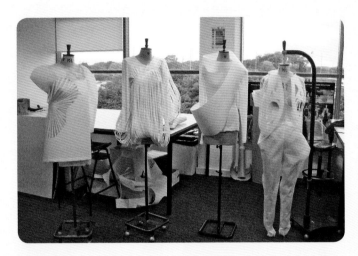

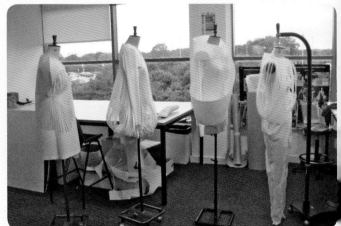

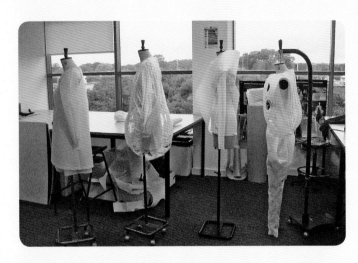

组合样衣审样

制作的系列成衣摆放一起审视,有助于从细节到格调的调整,使个性统一,整体风格呈现。

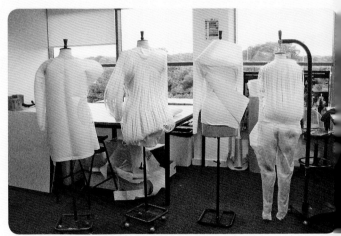

款式结构图
Flat Drawing

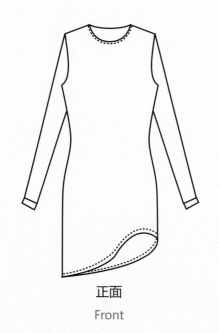

正面
Front

反面
Back

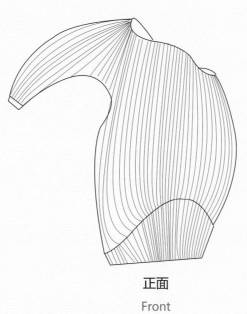

正面
Front

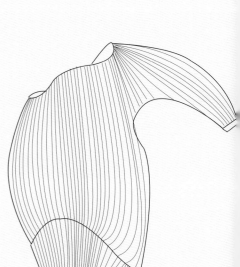

反面
Back

1号服装图例是简约感的大面积切割线裙装服饰

NO.1 garments: cutting line in big area of garments with simple sleeve

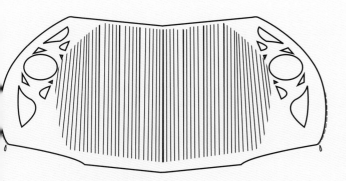
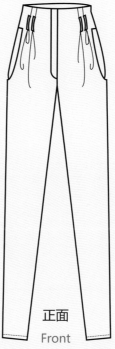
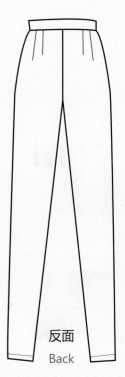

2号服装则着重强调了人体工程学对线条张力的作用及孔洞、弧线等元素的夹克设计

No.2 garment: cutting line in shoulder with hole element in Jacket

正面 Front　　反面 Back

款式结构图 / Flat drawing　　081

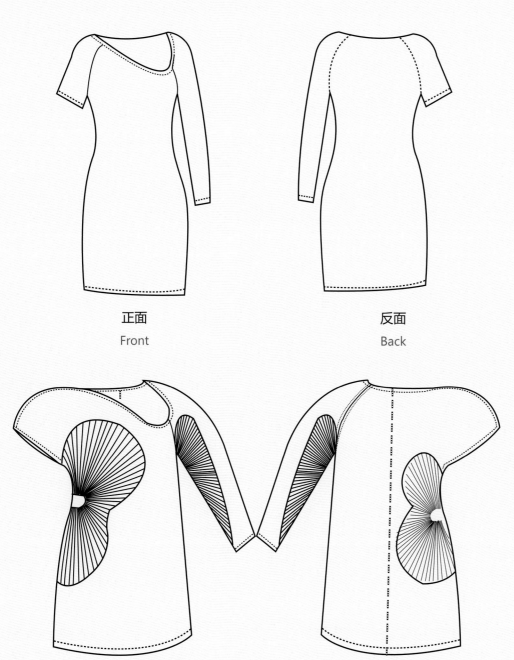

正面
Front

反面
Back

3号服装注重的是平衡不对称的服装结构和细节的裙装设计
NO.3 garment: balance of asymmetrical shape and detail

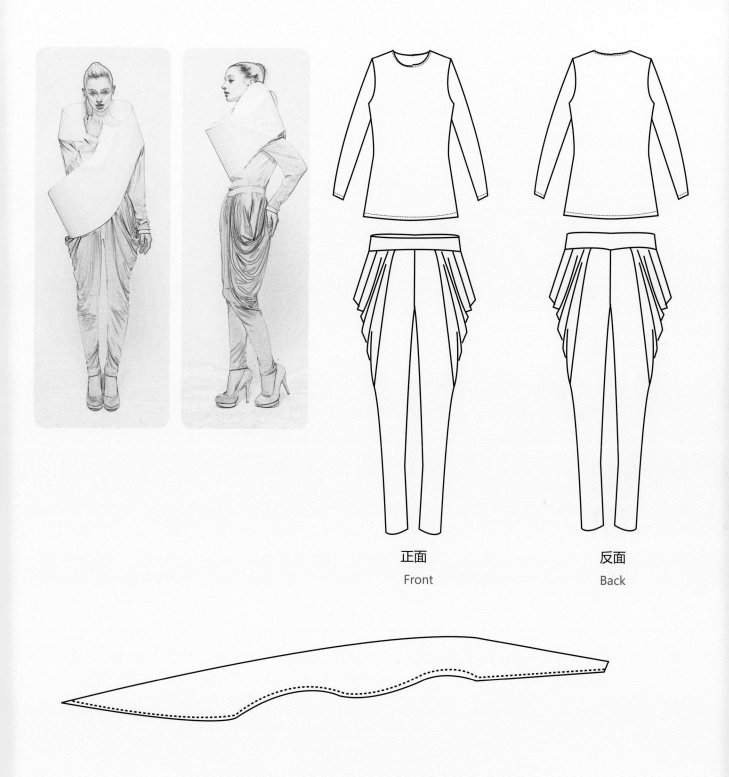

正面　　　　　　　反面
Front　　　　　　 Back

第4号服装是一个强调简约线条感的服装软雕塑上装设计。
No. 4 garment: simple roll top with line feeling legging

服装成衣搭配决策

将所设计的作品用实际面料加以制作,并按照前面所述的最终面料搭配方案搭配出来,看整体效果,如发现搭配面料有不当之处,给予更换,使之体现设计意图。

不对称线裁连衣裙
The Asymmetric Cutting Lines Dress

服装成品搭配效果及细节

一片式立体成型变化夹克三种搭配
The Three Changes of Cutting Line Jacket

服装搭配决策 /Decision making

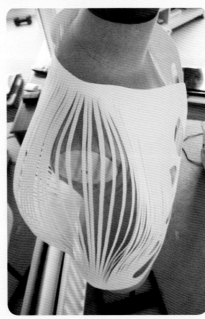

服装成品搭配效果及细节

雕塑感不对称袖子连衣裙
The Asymmetric Sleeve Sculpture Style Dress

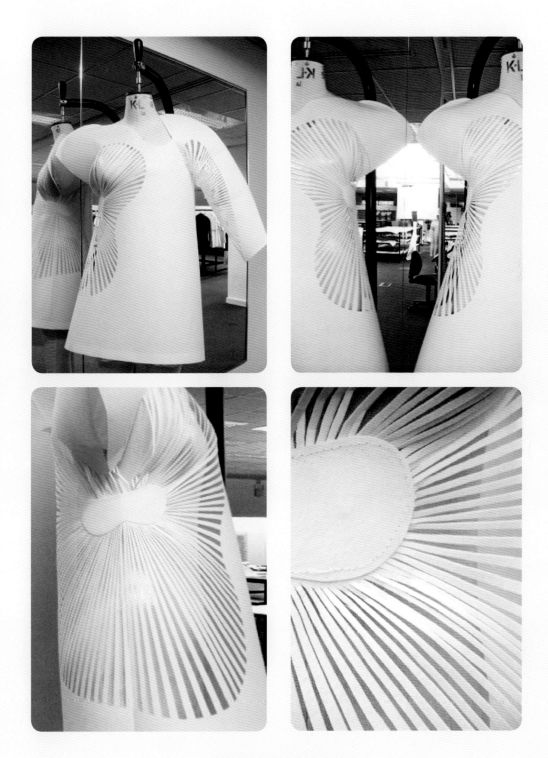

服装成品搭配效果及细节

旋转雕塑感一片立体成型变化上衣两种搭配
The Two Changes of Sculptural Style Roll Top

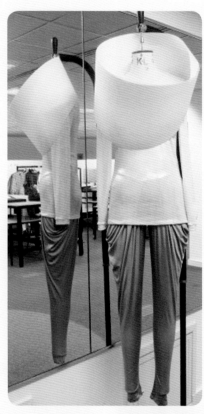

服装成品搭配效果及细节

饰品及其细节
Accessories and Details

配饰成品效果及细节

时装插画类似时装效果图,通过描绘展示出服装款式创意,在描绘中可在客观形状的基础上进行主观抒发,以充分表达作者的艺术感知和设计理念,使作品内涵浸透出浓郁的艺术感染力。

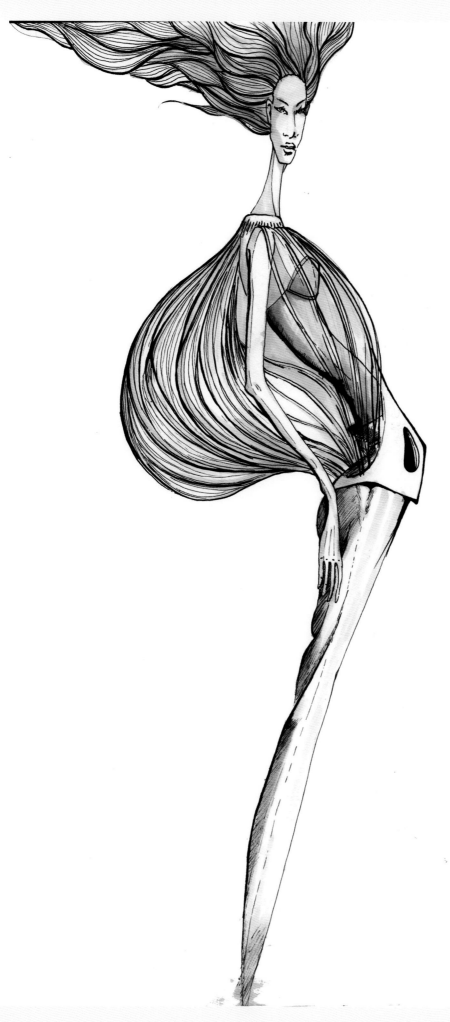

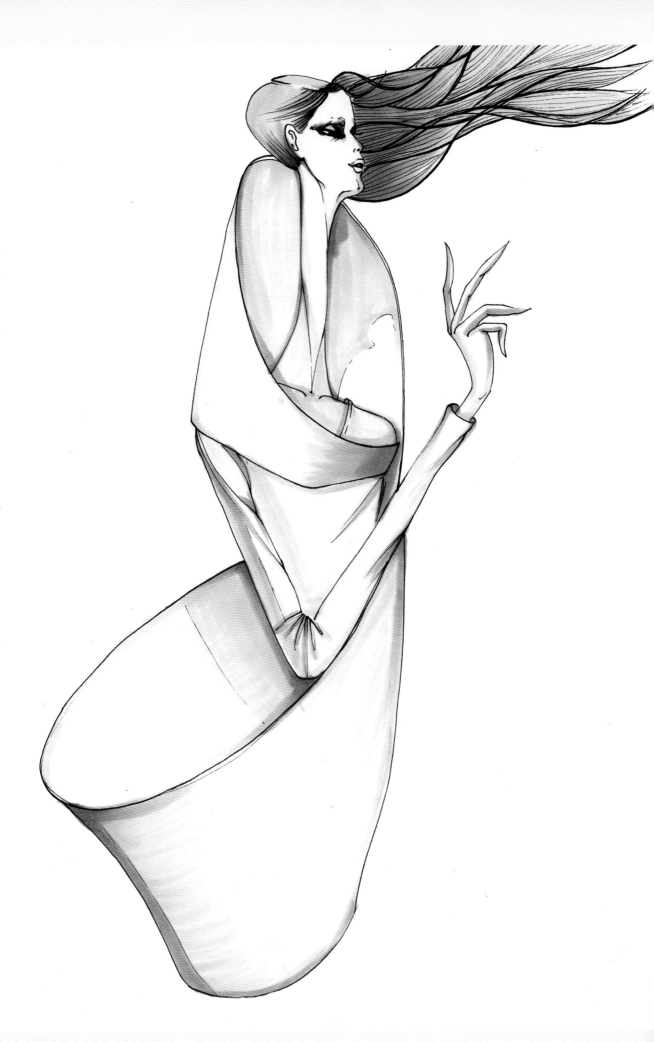

时装插画 / Fashion illustration

时装摄影造型设计
Styling Design Selection

时装设计师的创意作品及其美学艺术理念需要通过一定的摄影语言和模特造型才能充分展示，这一环节的内容构思也是整体创作重要部分，不容忽视。

整体摄影风格设计
The Photo Shoot style Design

构思灵感来源于 Exception 品牌的广告画,该画面虚与实的处理非常巧妙,空间人体给人以无限的遐想。

单个及组合摄影风格设定
The Each Photo Shoot Design

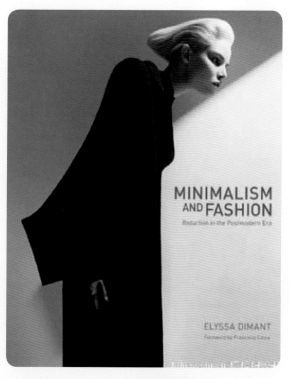

选择一些单个及组合摄影图像作为参考模型，这样有助于与摄影师进行沟通，有助于摄影师理解造型要求以及设计师所要表达的意图。

单体服装摄影风格设定：

黑白风格 / 用现代的手法体现服装雕塑感

组合服装风格设定：

    清晰

    现代感

模特选择
Model Selection

模特的选择非常重要。理解力，表现力以及身材气质具佳的模特才能演绎出设计师倾注在作品中的"精、气、神"。以下是两个待选模特的具体资料。

Evija：

身高：5'7" - 170cm

胸围：30 - 76cm

罩杯；B

腰围：24 – 61cm

臀围：34 – 86cm

鞋码：5 / 38

发色：暗金色

眼珠色：蓝

Grace Hollows

身高：5'9" – 175cm

胸围：31 – 80cm

罩杯；B

腰围：24 – 61cm

臀围：34 – 86cm

鞋码：5.5 / 38 / 6.5

发色：浅褐色

眼珠色：蓝绿色

Evija

Height: 5'7" - 170cm

Bust: 30 - 76cm

Bra; B

Waist: 24 – 61cm

Hips: 34 – 86cm

Shoes W: 5 / 38

Hair color: dark blonde

Eyes color: blue

Grace Hollows

Height: 5'9" – 175cm

Bust: 31 – 80cm

Bra; B

Waist: 24 – 61cm

Hips: 34 – 86cm

Shoes W: 5.5 / 38 / 6.5

Hair color: light brown

Eyes color: blue-Green

发型妆容设定
The Set of Hair and Makeup

总体妆容发型设定

模特发型妆容也需要采用一些专业图像来与化妆师沟通。

主体妆容要求：
  清晰
  简约
  唇膏用淡色系
  略微粗的眉毛

Makeup main look:
  overall clean
  Minimal
  Slight color fills on lids
  Stronger brows

主体发型设定：
  光滑干净
  略带雕塑感

Hair main look:
  Slick clean
  Sculpted in subtle modern way

时装摄影造型设计 /Styling Design and selection

模特拍摄姿势设计
The Designs of Photo-pose

构思草绘模特造型姿态提供给模特和摄影师,有利于他们理解设计师意图,并展开艺术再创造。

试装模特
The Fitting Model

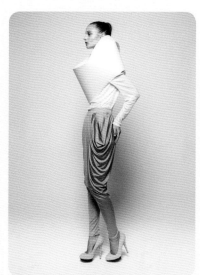

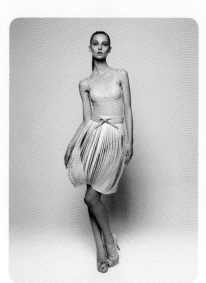

通过模特试装比较，决定最终模特人选，并演练全部艺术造型。

时装摄影造型设计 /Styling Design and selection

时装摄影

Fashion Shoot

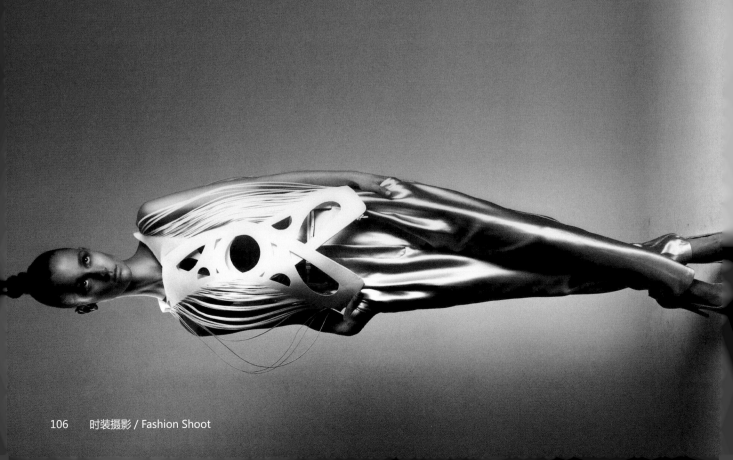

时装摄影 / Fashion Shoot

时装摄影 / Fashion Shoot

My Agenda for the Future
未来的工作研究展望

本次的硕士研究生毕业设计已经完成，但是这仅仅是时尚之旅的开始，而不是结束。通过这个设计项目的研究，本人研究分析了英国著名抽象雕塑大师赫普沃斯女士的作品，从中引发了极多的灵感和创新，并将雕塑空间中实与虚的平衡关系导入到服装设计当中。虽然，这次的毕业设计已经告一段落，但是本人意识到还有很多有趣的设计领域还未开发，值得更深入的研究。而对这些有趣设计领域的研究开发，就是本人今后的工作研究目标。

本人认为，随着时代的发展，艺术与服装之间的交融会变得越来越紧密。在过去的20年间，我们就可以看到时尚与艺术之间的跨界合作。很多设计师都纷纷与画家，雕塑家合作，设计出许多打破传统时尚观念和现代审美意识的新型时装产品。甚至许多大的服装设计品牌，如巴黎世家，普拉达和迪奥等推出的产品中可以看出：艺术世界给予了时装设计极大的设计灵感和设计借鉴。举个例子，巴黎世家2011-2012年秋/冬季时装秀上，设计师尼古拉·盖斯基埃就是从"新造型主义"艺术中得到灵感，玩转了时装的结构与色块的比例，很好的演绎出了"新造型主义"的精髓。

在未来，本人希望对抽象艺术对现代设计的影响做进一步的研究。例如，抽象艺术中对点线面的运用；简约风格中颜色节奏的把握；特别是，如何运用二维的几何形状去创造三维的空间感。

所以，第一点，在未来研究或工作中，本人会把研究重心进一步放在人体与服装空间实与虚的关系上。这主要是因为本人一直对这一领域非常的感兴趣，因此想进一步的拓展探究这一领域的研究。

与此同时，审视内心，本人意识到对于如何平衡抽象艺术的比例和细节，本人同样具有强烈的好奇心。（奎因，2003年；沃克，2011年；迪麦特，2010年）本人认为深入的去探索和研究这一个领域，对于成衣设计师来说，将会有莫大的帮助。当然，其他相关领域的研究我也感兴趣。比如，如何将抽象艺术中的结构运用到服装设计当中；如何从蒙德里安的绘画比例细节中获取创意，并运用

The design for my MA course has finished, but this journey of fashion design has not finished yet; instead it is just a start. By doing this project, I have researched the idea of solid shape and negative space in abstract sculpture (Hepworth, 1956) which has inspired me a great deal for this project, but I think a lot of other fields that need to be researched in a great in-depth for my future fashion design agenda.

I think the garment and art have always been closely combined, especially in the past 20 years; meanwhile, the fashion and art have been more and more in cooperation. A lot of designers get a number of ideas from the painting or sculpture to create the fashion, which breaks the traditional garment fashion and develops the concept of modern aesthetic. Even in the last 10 years, the big famous brands like Balenciaga, Channel, Prada and Dior have also launched new products, which transformed different types of art into the fashion. For example, In Balenciaga Fall / Winter 2011-2012 fashion show, the designer Nicolas Ghesquiere transferred the Neo-Plasticism art and play the proportions which inspired by the Neo-Plasticism art.

Thus, in the future, I would like to conduct further research in investigating how the abstract art impacts on the modern design, like the use of point, line and plane in abstract art; the rhythm of colour and simplistic style; especially, how to use 2D geometrical form to create 3D space.

Thus, first of all, in my future work or study, the key is to develop the research of solid shape and negative space in the application of fashion design. This is because I have always been interested in the concept of space between the human body and garment. Therefore, I will develop this concept

巴黎世家 2011-2012 年秋/冬季时装广告大片
Balenciaga Fall/Winter2011-2012 fashion show

山本耀司 2009 年秋冬秀中红与黑羊毛外套
The red and black boiled wool jacket and skirt in Yohji Yamamoto Autumn/winter 2009

到服装外观设计当中；以及如何用摩登的服装语言演绎几何抽象艺术。例如，"构成主义"的抽象艺术家们，强调空间中的势 (movement)，而不是传统雕塑着重的体积感。将传统雕塑的加和减，变成组构和结合，并结合一些科学的分析方法去支撑他们的艺术思维，构筑出了伟大的作品。本人认为这些思维可以运用到当代成衣服装的设计中来，比如分析衣服的实体廓形，装饰或结构，并在其上面做加法或减法，来形成新的服装廓形、比例或结构。就像是山本耀司2009年秋冬秀中红与黑羊毛外套与裙子一样。（萨拉萨尔／山本耀司，2011）。

最后，不断思考着如何将所研究的内容跟中国的传统时装设计结合在一起，创造出新的时尚概念。这个概念的产生也是通过对日本设计师的设计理念研究得出的，如川久保玲、山本耀司、三宅一生等，将日本和服和"禅意"与现代服饰相结合，诞生了属于日本的美感服装。

总而言之，通过这些设计与研究，为中国的时装事业做出一点贡献。本人会深入的探索和发展本人的一些想法，然后一步步实现它。我更希望，通过我的设计和研究，能够启发和引导更多的设计师深入这个领域，进行进一步的创新开发。

further and design a collection of garment to explore this idea.

In the meanwhile, I have realised that I have a great deal of curiosity about the balance of proportions and details in Abstract Art (Quinn, 2003; Walker, 2011; Dimant, 2010). I believe that studying and researching this area will also help a designer to design ready-to-wear garment. What is more, I will study more on other topics I am really interested, like how to use the balance of structure in constructivism abstraction in fashion; how to get inspiration from Mondrian's painting proportion detail and use it in garment's overall design, as well as how to use modern mind to transfer abstract Geometric style to fashion garment language. For example, the artists of "Constructivism" abstraction show their idea by reducing some unnecessary ingredients supported by structural rheology, human body mechanics and geometry. I think this idea can be used in fashion area, maybe I can analyse the figure of garments by solid geometry, quantification of volume, and other analysis to change the main structure of the garment, just like the red and black boiled wool jacket and skirt in Yohji Yamamoto Autumn/winter 2009 (Salazar & Yammoto, 2011).

Finally, I find myself continually thinking about how to combine my research with Chinese traditional garment together, in order to create a new concept of fashion. This idea is from Japanese designers: Rei Kawakubo, Yohji Yamamoto and Issey Miyake. The common feature of their garment is mixing the heavy traditional garments with Japanese concept of wabi-sabi (incomplete aesthetics), based on which they have developed and created their current modern garments.

论文：空间的遐想 ——将雕塑艺术导入服装设计

The Imagination Of Space
-- Transforming Sculpture into Fashion

设计灵感源及研究领域

本人的硕士研究生毕业时装设计作品重在于对实与虚关系的探讨，是对服装空间的一次大胆的遐想，也是一次非常有趣的探索性旅程。

作品的设计深受英国著名抽象艺术大师赫普沃斯女士艺术思想及其雕塑作品的启发（Hepworth，1956），人体与服装的空间关系是研究探讨的主题。设计构思的重点是服装结构实与虚的平衡关系，作品效果舒展流畅，端庄典雅，灵动美妙。探索过程思维开放，步骤清晰，结构严谨，新颖创新。

总之，该系列作品的整体设计思路是探讨人体与服装的空间关系。设计的创作构思是基于对赫普沃斯女士抽象雕塑的研究，以及对其他具有相同理念的设计师作品的探索及借鉴之上的。主旨是将抽象雕塑艺术导入服装设计，以拓展时装设计的空间领域。

赫普沃斯抽象雕塑的启发

本人是从赫普沃斯的抽象雕塑并研究分析其艺术特征开始这次毕业设计之旅的。研究的展开从一些小的细节开始，例如，分析赫普沃斯的抽象艺术特征；以及研究赫普的雕塑是如何平衡结构与细节。这些细小却又重要的特征细节非常重要，往往正是这些微小的细部特征，引发设计师们无限的创意。

在研究沃斯女士的作品时，本人深深的被她的雕塑所吸引，它使人们联想到宁静而优雅的停歇在湖畔的天鹅；或是双臂舒展，在幽静的练舞房内翩翩起舞的芭蕾舞女演员。它是如此的令人陶醉着迷，引人浮想联翩。

从赫普沃斯（1943年，1946年，1956年）的作品中，可以发现流动优雅是她造型艺术的一大特征，这种独特的造型使得她的雕塑散发着典雅迷人的韵味，从而触发本人无限的灵感和遐想。比如：从赫普沃斯的雕塑作品中，本人发现将雕塑实体造型做得光滑圆顺，结合自然或抽象的

The practitioners who have influences on my design

The project of my fashion design concentrates on solid shape and negative spaces. In this project, I designed a series of the ready-to-wear garment that explore the space between the human body and garment, this idea of which was influenced by Barbara Hepworth's abstract sculpture (Hepworth, 1956). In particularly, I focused on the balance of solid structure and the area that were not solid, in order to create the garment to have clear silhouette and elegant detail. Therefore, this section will present the ideas that my design was based on Barbara Hepworth's abstract sculpture and other designers' ideas.

The influence of Hepworth's abstract sculpture on my design

I started with researching Barbara Hepworth's abstract sculpture and analysing the characteristic of Hepworth's art, which includes the figurative art, and the balance of structure and detail in Hepworth's sculpture. I was attracted by Hepworth's sculpture, which reminded me of an elegant swan or a ballet dancer, who was stretching out their bodies in quiet, minimalist and elegant. They were fascinating to me.

From Hepworth's (1943, 1946, 1956) work, I found the figurative art is a strong character of her work, which made her sculpture very elegant, and inspired me to employ figurative art in the design of my garment at a later stage.

In her sculpture, I also found she liked to highlight the solid shape which she made very smooth and combine the abstract and/or naturalistic characteristics in the convex or concave of curves, which is another characteristic of Hepworth's sculpture. Hepworth's sculpture usually has geome

凹凸曲线，塑造出的雕塑作品，是极具有抽象艺术独特魅力的。她的雕塑通常由简单的几何形状组成，具有高度抛光表面和流畅优雅的曲线。她坚信将雕塑的实体部分设计成优美的曲线或造型是十分必要，因为这将会使观众对雕塑的外观留下深刻的不可磨灭的印象（赫普沃斯，1956年）。

其次，更值得一提的是在对沃斯女士的雕塑进行进一步的研究时，本人还发现她喜欢运用不同尺寸的孔洞来贯穿整体雕塑，形成一种特别但美观的雕塑结构，使得空气，空间和光线得以在这些孔洞里交相辉映，形成雕塑实体内的第二个虚拟的空间。这样的结构设计，空间的贯穿给予了传统雕塑新的生命和崭新的活力。

基于孔洞的设计，她雕塑的另一个显著的造型手法是对线条的运用。线条的穿透力和连接性可以很容易的使雕塑产生灵动的空间感，并且能够平衡实与虚之间的空间关系。例如，雕塑作品"木材与弦线的'赋格'"就将空间的力量与平衡处理得非常完美。雕塑品"木材与弦线的'赋格'"是由一系列的抽象几何形体组成：强有力的线条和优美圆滑的孔洞巧妙的组合在一起，共同谱写出了了由木头和弦线组成的节奏感很强的乐章。从这个雕塑中，人们可以看到有不同的设计线条从雕塑造型的一边紧密的联系着边缘光滑呈优美弧线的另外一边。"线"在这个雕塑作品中就像是一座桥梁，将两个不相关联，而且，这个线不是普通的连接，它是经过周密设计的。沃斯女士（1962年）将其设计成了美丽的三维立体的网状，从不同的角度去看这个网状的"线条"的布局，欣赏者会随着探索的深入，观察手法的不同发现更多的趣味来。真是步步是设计，点点是细节。

正是这样的散发独特魅力的造型以及巧妙设计的细节深深吸引着本人的眼光，使人产生无限的遐想空间。并启发了强烈的创作欲望。特别是通过"线条"与"孔洞"来平衡空间虚实的关系的触动，使本人产生了将其运用到时装设计中去的创意。对此本人展开了一系列的创作构思，草图勾画和艺术的联想。

-trically simple shape, which has highly polished surface with elegant curves and a variety of holes of different sizes. She believes that it is necessary to develop a sense of beauty for solid shape, or "form", because it will improve the appearance of the sculpture and impress the audience (Hepworth, 1956).

What is more, she also liked using holes with different sizes in her sculpture to enlighten the structure for air, space and light to get through. Those holes gave breathing to her sculpture (Director, 1982). Besides using holes, another strong characteristic of her sculpture is the use of lines. The lines are very important for making sense of space and making the balance of solid shape and the area that is not solid. For example, by observing the sculpture Wood and Strings 'Fugue', I perceived a combination of balance, strength and serenity of this sculpture that I was appealing to me. The sculpture 'Fugue' is composed of a series of shape: strong lines and holes which combine together and then compose as a completed one piece of art. From this sculpture, people can see different designed lines from this sculpture, people can see different designed lines from one side of the sculpture, and smooth and curving edges of the wood from the other side of the sculpture. The lines in 'Fugue' look like a bridge connects one side to another. In addition, Hepworth (1962) has also made the lines as beautiful as a spider web.

Hepworth's sculpture impressed me and inspired me for my own fashion design. I was impressed by the use of lines and holes to make the sense of art, and also employed the idea of using lines and holes in the design of my garment. In the application of the lines and holes in my fashion design, the lines can not only connect two blocks together, but also leave little space for people to see the shape of

例如，本人联想到"线"在本次时装设计中，其作用可以不仅仅是联系左右两块不同衣片，还可以通过线与线之间的交错变幻，形成一种随人体运动而变幻的设计面。"孔洞'也可以不仅仅是装饰的作用，人们通过它还可以欣赏到"第三空间"——也就是建立在人体自然曲线与服装廓形之间的空间美。设计师还将运用几何造型与不规则造型交织的强烈视觉感来设计这组成衣的具体廓形。这也就是本次毕业设计的主旨思想——创新与平衡——将赫普沃斯独特的艺术抽象美感导入服装设计当中，使时装也散发出雕塑艺术的魅力。

相同领域服装设计师们的研究与启发

在调研完赫普沃斯 1943 年 -1958 年的抽象雕塑作品后，通过对许多当代的著名时装设计大师的研究，例如：川久保玲，三宅一生，三本耀司和候塞因·卡拉扬。他们也给予了本次设计思路许多新的启发。

这些设计师都是长期专注于服装廓形和纺织面料的研究，并将艺术与服装产品完美的融合在了一起。比如像川久保玲。川久保玲是当今世界最有影响力的日本设计师之一。她改变了 20 世纪末 21 世纪初人们的穿衣哲学。她创造出了新的创意时尚——超大尺寸，信封式的廓形。她改变的不仅仅是传统的女性时装廓形，她还引导当代女性进行无彩色，无性别的着装革命。她的设计专注于研究人体与服装的关系，比如在女性的肩部和臀部做夸张放大的设计，这与传统的优雅高贵的法国高级时装廓形相去甚远。（清水，2005 年）但是，也正是这样的大胆创新，使她的作品引导人们对人体本身曲线的思考。

本人对她的 Comme des Garcons 品牌在 1997 年推出的春夏时装 "身体偶遇裙子，裙子邂逅身体" 主题系列非常的感兴趣，由此展开了一系列的分析。其实，但看这个系列的名字就能让人浮想联翩，会使欣赏者不禁对一件裙子如何的与人体邂逅及亲密接触的过程产生一系列的联系。这组系列用紧身弹力材质与可拆卸衬垫的塑造了一系列独特有趣的时装造型。在这一系列中，川久保玲将这些紧身弹力材质包裹可拆卸衬垫设计放入服装的肩部、背部，身侧和臀部，从而缔造出了一种人们从未见过，夸张

human body. The holes can be used for people to see the 'third space' which is the space between the garment and human body in my design. I will also use strong silhouette geometrically and asymmetrically in the parts of garments. In the meanwhile, I will use curving silhouette, lines and holes to create and balance the shape of outside abstract garments with the outline of inside human naturalistic figure, which is the main concept employed in my fashion design.

Fashion designers' impact on my design

Besides the influence of Hepworth's (1943, 1946, 1956 1958) sculpture, my fashion design was also influenced by others designers, such as Rei Kawakubo, Issey Miyake, Yohji Yomamoto and Hussein Chalayan.

There are designers who are well known for taking an artistic less commercial approach to fashion in term of silhouette and textile, such as Rei Kawakubo. Rei Kawakubo is one of the world's most influential fashion designers. She changed the mind of people who lived in the 20th century. She found the new way to create fashion by designing oversized garment that envelope the female form rather than traditional, exposing and non-colour garment. Her design liked to investigate the relationship of human body with garment, such as exaggerating the part of shoulders and hips which was far from the heavy traditional French high fashion (Shimizu, 2005).

I am really interested in her Comme des Garcons Spring / summer 1997 collection; the name of this collection is "Body meets Dress, Dress meets Body". The collection name is powerful and makes me think about the nature of physical beauty and femininity. Her collection includes a range of figure-hugging stretch garment, with removable down pads sewn inside the garment. Rei Kawakubo's pads were

到极致的但又无比惊艳的廓形剪影。正是这组川久保玲的设计，使本人也产生了设计"当身体邂逅时装"的创作灵感。

在完成了对川久保玲的研究后，接着，本人对那些在处理服装结构与细节的平衡关系上的设计大师进行深入研究，比如三宅一生．

三宅一生著名的可塑性褶裥织物。这个褶裥织物是建立在人体结构上，并可以塑造出不规则的几何轮廓。三宅一生的时装设计与传统方式完全不同。他将剪裁的焦点放在了如何建立服装的结构和造型的比例上，以刻画一种"禅意"的织物雕塑感，被誉为最具有神圣东方哲学韵味的时装设计。除此之外，他还关注服装造型与细节的平衡（三宅一生，2010）。这在三宅一生早期作品中，尤其显著。他所塑造的一系列抽象几何褶皱服装，例如1994年春／夏季系列的"飞碟礼服"，就是用几何剪裁切割与皱褶的特性，去塑造一个新的、完全打破传统规则的廓形，而且这个廓形会随着人体的运动方式产生不一样的变化。甚至于，人们在不同的角度欣赏同一件衣服，它所产生的廓形美感也会不同。（布莱克，2006a，2006b）

另一组带来强烈灵感冲动的三宅一生的作品是：2010年三宅一生的"1325"主题系列———这个系列以折纸手工和可持续性理念为基础。该系列名称中的数字1是指每件衣服由一块布料做成，3是指衣服呈三维立体形状，2是指衣服可以被折成二维形状，5是指每件衣服有多种穿法。（三宅一生，2010）它是通过人体的体型来撑出衣服的三维立体的廓形，本体的服装其实只是一片经过独特设计的面料。 这个概念触动了本人很好的设计思考：如何运用面料所本固有的特性，设计出建立在人体结构学上的时尚服装？

与三宅一生设计理念相似的日本设计师是山本耀司，他被誉为日本时尚界的教父，他同样在时装设计中注重服装结构与细节的平衡关系。

他的服装有其独特的抽象感，特别是他服装设计中的对称

inserted at the upper back, side and shoulders, which creates extreme silhouettes, the likes of which had never been seen on catwalk (Fukai, et al, 2010a). Rei Kawkubo's designs inspired me to research and create the relationship of "Body with dress" in my own design.

Besides Rei Kawkubo, I take a close looking at the designers who focused on the balance of structure and details, such as Issey Miyake.

I was inspired by Issey Miyake who was famous for using pleating fabric to form geometric outline, which was built in human structure. Issey Miyake's design ignored traditional cutting, but focused on the proportions between clothes and figurative sculpture. In addition, he was also successful in balancing garment figure and fabric detail (Miyake, 2010). For example, in Issey Miyake's early collection, he created a series of dynamic garment by using abstract geometric form with pleats fabric, such as "Flying Saucer dress" in Spring / Summer 1994. The garment used geometric cutting and the characteristic of pleats to create an abstract form, in order to create different shapes whilst movement and be appreciated at different points of view (Black, 2006a; 2006b).

Another example of Issey Miyake's creative design is 'A piece of flat material become a three-dimensional structure' in his "1325"collection, 2010. The structure in turn can become a two-dimensional shape with the addition of straight lines. But when human wear this piece of flat material, it can be a new different shape based on the human body shape (Frankel, 1997; Miyake, 2010). This also inspired me in my own garment design in terms of how the garment can build in the human body structure; while taking maximum advantage of the fabric's own feature.

感与不对称剪裁的平衡关系非常的令人着迷。这种特殊的对称性在他的时装中也不是传统意义上的对称感，而是用不对称的裁剪设计出一种人类视觉上的对称平衡。例如：在 2001-2002 阿迪达斯／山本耀司秋冬发布会中，他所设计的经典 Y-3 条纹海军蓝羊毛华达呢拉链外套就充分体现出了他用不平衡来平衡视觉的设计思路。衣服右侧设计成由经典 Y-3 条纹装饰的完全束缚不可动的大袖子，而另一侧则设计成深蓝色简单的可灵活活动的紧到极致的左袖。他用不平衡的设计——用紧而灵活的左袖去平衡松而束缚的右袖，又用大面积但深沉的蓝色去碰撞小面积但活力四射的条纹——在视觉上反而塑造出一种惊人的和谐感、平衡感。

除了用不对称的剪裁设计来达到视觉上的平衡，"简约"是三本耀司时装设计的另一个显著的特点。他的时装都有一种独特的设计美学，叫做"残缺美"。因此，你在他的作品中会发现，他摒弃了一切其他设计师增添的繁复细节，使他的作品散发出"简约"的魅力。此外，他服装的简约雕塑感是完全建立在人体工程学上的，这点可以从 1996/1997 年山本耀司秋冬秀中的白色毡步连衣裙中得到反映。这件裙子，他重点突出了礼服的背面。他用精湛的剪裁和精致的细节处理服装背部的结构，从而凸显了裙子前片的简约感。而裙子后背的开口，结构线的弧度处理，面料材质的运用，面料如何在人体上站立的方式等等，都是建立在设计对人体工程学的理解上。使看似简单的时装作品散发出雕塑才具有的艺术气息。以上这些也正是我从对山本耀司的作品中所感悟的，即建立在人体结构上的简约，和用不对称的剪裁设计来达到视觉上的平衡。

在服装材料的研究上，本人也调查研究分析了一些善于利用材料特性或善于运用特殊方式使用服装面料的服装设计师，比如像是候塞因·卡拉扬．

从侯赛因·卡拉扬历届的时装秀中的作品，可以发现他在服装设计中，非纺织类材料运用之多，材料的选择范围之广，创意的点子之奇，是一般设计师难以比拟的。例如，在他的 2000 年春夏时装发布会上的"飞机"礼服的材质就是由玻璃纤维压膜制成的，玻璃纤维立体成型形成了光

Similar to Issey Miyake, Yohji Yamamoto is another famous designer focuses on the balance of structure and details.

His clothes are in a scene of peculiarly abstract. The way how he balances the asymmetry and symmetry in his design fascinated me. Symmetry in his garment is not in a traditional definition; in fact, the symmetry is reflected by his application of asymmetry to make a visual balance of the garment. For example, in Adidas /Yohji Yamamoto Autumn/Winter 2001-02, he designed the Navy-blue Wool Zip-opening Gabardine Jacket which has a white striped inlay in the right sleeve which is restricted with the body jacket. Other parts of the garment are totally in blue colour with a tight left sleeve. The balance of symmetry and asymmetry reflects on the colour and sleeves. In terms of the colour, the large area of blue balances the small area of white strips. With regard to the two sleeves, the tight left sleeve balances the overall constrained feeling of the garment caused by the restricted right sleeve (Fukai et al, 2010b).

Besides the feature of a balance of symmetry and asymmetry, simplicity is another feature of Yamamoto's design. His design employs the concept of 'aesthetics of the incomplete', so he cut off all the unnecessary details rather than adding details as some other designers do. Additionally, his sculptural form is based on human being's natural structure, which is reflected on the Yohji Yamamot Autumn/Winter 1996-97 White Wool Felt Dress. Of that dress he emphasises more on the back of the dress where he used cutting and detailing to balance the simple front of the dress (Washida, 2002). Therefore Yamamot's design taught me how to create a balance and how to keep simplicity in my own fashion design.

吉而简约的廓形，变成一个雕塑般独立的"身体"。在这光洁的玻璃纤维表面下，是有网眼纱层层叠叠组成的蓬裙。当模特展示时，雕塑般的表面在高科技的控制下慢慢升起，里面的网纱蓬裙便柔美的演绎浪漫的风情（奎因，2003：67）。从侯赛因·卡拉扬的设计中，又引发了本人许多灵感，设想在毕业作品系列中运用一些意想不到的材质来设计雕塑感般的造型。但是如果选择这样的方式，那么，如何去缝制这种非常规材质就会成为一个新的挑战。对此，本人更加深入的对侯赛因的缝制的处理方式，进行了相应的研究。研究结果表明，由于侯赛因几乎是运用硬质材质的材料进行设计，他一般是将两种材质进行不同的处理。纺织类面料还是遵循常规的缝制手法；非常规面料则会动用焊接，压膜等高科技手段。这一系列的缝纫研究给予本人许多启发，那就是将不同材质，分别用不同的拼贴手法处理。不过，其中非常规类材质的处理方式，并不符合整体的时装设计宗旨。因为本人的设计目标是设计制作出一组高级成衣系列，那么过于非常规的处理方式可能在成衣系列中并不是那么的符合成衣消费市场。因此，非常规类材质如何与普通面料进行缝合，会在之后的实验中成为最为重要的实验项目。

本次高级成衣时装设计实验过程

基于对上述对艺术家及相关设计师作品的调查研究，接下来，本人针对高级成衣目标客户群做了市场调查。本次高级成衣的目标客户群的年龄层是28-38岁的时尚成功女性。根据ERAY高级营销公司2009年-2010年的市场调查报告中分析，针对这类女性的服装色彩，面料需求，款式喜好，主要社交场所等特征进行了系统的分析。数据表明28-38的时尚成功女性，主要社交场所是一些正式的中高端场所，比如像是展览或鸡尾酒会。着装颜色偏向于细腻和淡雅的色彩，如白色，浅灰色，裸体和黑色。面料选择则倾向于舒适，自然和高品质的面料，如100%桑蚕丝，100%纯棉弹力面料。设计款式风格则多为简约、优雅的时尚设计。

根据以上对灵感源以及市场的调查研究，最终，本人设计出了八套强调实与虚空间感，并具有简约优雅线条的高

In terms of the material, there are designers who combine traditional and modern methods to make good use of the special materials. The typical one is Hussein Chalayan.

In a collection of Hussein Chalayan's fashion work, I found that he used a broad range of unusual materials to design garments. For instance, in his Spring/Summer 2000 show, "the 'aeroplane' dress was made of moulded fibreglass, with moving panels revealing a froth of tulle. The fibreglass panels of this dress form a dense structure that maintains its shape independently of the body." (Quinn, 2003: 67). Inspired from his design, I thought my collection could use some unexpected materials to make the shape of garments, but how to sew these materials could be a challenge. Chalayan usually screwed the wood and other hard materials together, which would not be applicable to my fashion design because my aim is to create a ready-to-wear collection. Therefore, the choice of materials needs to at the convenience of sewing methods.

The practice of my fashion design

Based on the ideas of other practitioners as discussed above, the target customers of my garment were accordingly targeted. Also referring to the marketing report of ERAY Company 2009-2010, my target customers were focused on the following attributes: age, venue, colour, fabric and style. The age of the target customers would be between 28 and 38 mature females; the occasions when they wear my garment would be for formal and smart functions, such as exhibition and cocktail party. Colour of the design needed to be delicate and subtle colour, such as white, light grey, nude and black to show its elegance. Comfortable, natural and high quality fabric, such as 100% silk , 100% cotton and stretch fabrics would reflect the elegance. The style needed be simple but considered design.

级成衣。由于时间的限制，最后，从中挑选出四件作为这次研究生毕业设计的最终制作主图，进行更加深入的优化，制作。

本次的设计是从对服装面料的选择开始的。在经过层层实验和筛选，本人择取了白色的重型无纺织面料来塑造整体的服装造型。重型无纺织面料不是常规的服装制作面料，它通常被用于制作工艺品、帽子坯样，或当粘合板给灯具做装饰。不过，本人选择它作为服装主打面料，是因为它能够提供足够的塑造性和韧性。

但是，在前期的塑形实验中，本人发现，这种面料虽然极容易塑形，轻易可以达到设计师所追求的雕塑感的廓形；但是，由于它并不属于常规性纺织类材料，材料硬度过大，并不适合贴身穿着，而且，它容易造成人体活动的不便，活动受限制。当本人尝试着将其裁剪成普通服装的结构，并进行着装活动实验，着装者的手臂在袖筒里不能自由弯曲，活动极其不便。

为了解决这问题，本人进行了更多的面料重塑实验，希望这类面料在满足我塑形要求的同时，又能满足人体活动的基本需求。首先，设计师对这种重型无纺织布进行褶皱处理，希望能将其变得柔软。但是结果并不理想。普通柔软面料在进行褶皱处理时所能达到的清晰的褶皱感，在这个面上得不到呈现；反而这种模糊的褶皱感破坏了材料本身的塑形优势，使造型变得凹凸不平，不具备赫普沃斯女士雕塑作品的圆润感。这不是本人所期许的。

褶皱实验失败后，本人马上针对所出现的问题，做了另一个面料重塑试验，试图解决目前所遇到的问题。在这次试验中，本人用一把美工刀将这种重型无纺织布切成一条条间距0.3cm宽，线条宽度也是0.5cm的细线，以此来消除面料的僵硬感。之所以采用0.5cm的宽窄的线条剪裁，是因为通过实验证明，0.25cm / 0.3cm的宽窄会使面料的可塑性大大降低，廓形的韧性也会下降，并且面料的着装牢固度也会降低。而采用0.7/0.8cm宽窄的剪裁，牢固度得到了加强，但是线条太过粗放，精细度不够，不符合赫普沃斯抽象雕塑的优雅感。而0.5cm宽度比例，最能完

Bearing in mind of the inspiration from influencing practitioners, I have designed 8 sets of ready-to-wear garments which have strong simple outline and mixed solid forms and negative spaces. Due to the time limitation of designing the garment, 4 designs were chosen to form a collection for this project design. I started with choosing materials for my design. I chose white heavy non-woven fabric to create the shape of my garment. This fabric is usually used for crafts, pelmets and hats, as well as interlinings or foundation, because it can provide specific functions, such as resilience and strength. Therefore, using this fabric would create and maintain the silhouette of my garments. However, during my experiments, I realised that though it was easy to make shapes with it, it was difficult to wear, because it was very inflexible. When I tried to wear it, my arms could not bend. In order to solve the problem and make it flexible where flexibility was required. I, firstly, pleated the material and made it softer; however, the result was still not suitable for my design, because the outline of my collection needed to be strong and clear. Pleating the fabric made the outline unstructured.

After that, I did another experiment with the fabric. I used a knife to cut the fabric line by line. The space between the lines is only 0.5 centimetres, which has an allowance between one line to another, which can also help the movement of human body. In addition, every line can make different shape when wearer moves because of the fabric's function included in the collection. In design wise, it is in keeping with the other cut work, the audiences can see the curve of human body through the allowance. The result of cutting lines in the fabric proved to be very successful.

After the moderation of fabric, I modified the draft design of the garments. I decided to use the line to become the main element of my garments' detail. Then, I needed to resolve how to integrate the cutting line into the overall look of the garments, in order to the balance of solid structure with details. From Yomamoto's garments (Quinn, 2003; Black, 2006a), I found his design detail is built based on the human body structure analysis, so the details appear in his design is not abrupt but looks very elegantly luxurious. Thus, I used the software (Rhino) to simulate the movement of the human body, to understand the changes in the structure of movement.

From this simulation, I found the joint of human body is to help human movement. Thus, I arranged cutting lines in the areas of shoulder, back, armpit and elbow and used tension of lines in these area to help with movement. Moreover, I reviewed Hepworth's sculpture 'Wood and Strings (fugue)1956' (The Tate Gallery, 1982; Thistlewood et al, 1996), and felt the sense of femininity, smooth and soft line definition, such as curve, roundness and holes which are characteristics of her work. Accordingly, I designed these elements in details of my garments such as neckline, garment bottom, sleeve and wrist.

For the space between the human body and garments, I designed the outer garment which is big and creates space between the garment and body figure. In this way, the audience can see the inside body curve through the allowance between the lines which will also give the audience an imagination of space.

Meanwhile, I kept focus on the balance of structure and details. When I used the cutting line in big area of garments, I kept simple design in neckline, sleeve and garment bottom. When I designed the cutting line in parts of garment, I also made other elements

廓形和紧身的基本内搭，来营造人体与衣服之间的空间感。欣赏者可以通过线与线之间的空隙，影影约约的欣赏着人体自然的身体曲线的美感（见图片25），而夸大的廓形，又可以给人们有一种雕塑的艺术美感，两者交错融汇，相辅相成，从而使人们产生一系列对服装空间的遐想。

当然，本人在设计构思中还一直注重服装结构与细节的协调。当本人大面积地在服装上进行"线条"设计时，会在颈线，袖子和衣摆等细节做简约的设计处理。当本人只是在服装的某个部分设计"线条"时，则会将其他诸如孔洞，弧线等元素放在一个会使视觉产生平衡的位置，避免服装在视觉效果上的失衡。

除此之外，本人还将不对称的剪裁运用到本次设计当中，但同时又使整体服装视觉上是平衡对称的。例如：在3号服装中，本人设计了的不对称的袖子。右边的长袖，本人将其设计的细而窄；左边的短袖，本人则运用夸张的设计手法，将其设计成圆弧状，高耸在左边肩膀上。与此同时，我又将线条细节运用在细长的袖子上，以此来平衡夸大的圆弧左袖所带来的视觉冲击；但又将中等面积的"线条"设计装饰在心脏部位，使3号服装的左右两边在视觉上产生平衡。通过运用这些细节上的处理，最终使不对称的剪裁在成衣设计中成为一种亮点而不是冲突。

最后，本人设计完善了四件一系列的高级成衣设计。

未来的工作研究目标

本次的硕士研究生毕业设计已经完成，但是这仅仅是时尚之旅的开始，而不是结束。通过这个设计项目的研究，本人研究分析了英国著名抽象雕塑大师赫普沃斯女士的作品，从中引发了极多的灵感和创新，并将雕塑空间中实与虚的平衡关系导入到服装设计当中。虽然，这次的毕业设计已经告一段落，但是本人意识到还有很多有趣的设计领域还未开发，值得更深入的研究。而对这些有趣设计领域的研究开发，就是本人今后的工作研究目标。

本人认为，随着时代的发展，艺术与服装之间的交融会

in neckline, garment bottom, sleeve and wrist for balance. Furthermore, my garments were designed asymmetrically; whilst maintaining the balance of a garment. For example, in the No. 3 garment, designed the asymmetrical sleeve, but I put the cutting line detail in the small sleeve side and used this detail to balance the other side of the big curve sleeve.

Finally, a collection of the four designs have finished

My agenda for the future

The design for my MA course has finished, but this journey of fashion design has not finished yet instead it is just a start. By doing this project, I have researched the idea of solid shape and negative space in abstract sculpture (Hepworth, 1956) which has inspired me a great deal for this project, but think a lot of other fields need to be researched in a great in-depth for my future fashion design agenda.

I think that the garment and art have always been closely combined, especially in the past 20 years meanwhile, the fashion and art have been more and more in cooperation. A lot of designers get a number of ideas from the painting or sculpture to create the fashion, which breaks the traditional garment fashion and develops the concept of modern aesthetic. Even in the last 10 years, the big famous brands like Balenciaga, Channel, Prada and Dior have also launched new products, which transferred different types of art to the fashion. For example, In Balenciaga Fall / Winter 2011-2012 fashion show, the designer Nicolas Ghesquiere transformed the Neo-Plasticism art and play the proportions which inspired by the Neo-Plasticism ar

Thus, in the future, I would like to conduct further research in investigating how the abstract ar impacts on the modern design, like the use of point

变得越来越紧密。在过去的20年间，我们就可以看到时尚与艺术之间的跨界合作。很多设计师都纷纷与画家，雕塑家合作，设计出许多打破传统时尚观念和现代审美意识的新型时装产品。甚至许多大的服装设计品牌，如巴黎世家，普拉达和迪奥等推出的产品中可以看出：艺术世界给予了时装设计极大的设计灵感和设计借鉴。举个例子，巴黎世家2011-2012年秋/冬季时装秀上，设计师尼古拉·盖斯基埃就是从"新造型主义"艺术中得到灵感，玩转了时装的结构/色块的比例，很好的演绎出了"新造型主义"的精髓。

在未来，本人希望对抽象艺术对现代设计的影响做进一步的研究。例如，抽象艺术中对点线面的运用；简约风格中颜色节奏的把握；特别是，如何运用二维的几何形状去创造三维的空间感。

所以，第一点，在未来研究或工作中，本人会把研究重心进一步放在人体与服装空间实与虚的关系上。这主要是因为本人一直对这一领域非常的感兴趣，因此想进一步的拓展探究这一领域的研究。与此同时，本人审视内心，意识到对于如何平衡抽象艺术的比例和细节，本人同样具有强烈的好奇心。（奎因，2003年；沃克，2011年；迪麦特，2010年）本人认为深入的去探索和研究这一个领域，对于成衣设计师来说，将会有莫大的帮助。当然，其他相关领域的研究我也感兴趣。比如，如何将抽象艺术中的结构运用到服装设计当中；如何从蒙德里安的绘画比例细节中获取创意，并运用到服装外观设计当中；以及如何用摩登的服装语言演绎几何抽象艺术。例如，"构成主义"的抽象艺术家们，强调空间中的势(movement)，而不是传统雕塑着重的体积量感。将传统雕塑的加和减，变成组构和结合，并结合一些科学的分析方法去支撑他们的艺术思维，构筑出了伟大的作品。本人认为这些思维可以运用到当代成衣服装的设计中来，比如分析衣服的实体廓形，装饰或结构，并在其上面做加法或减法，来形成新的服装廓形、比例或结构。就像是山本耀司2009年秋冬秀中红与黑羊毛外套与裙子一样。（萨拉萨尔/山本耀司，2011）。

line and plane in abstract art; the rhythm of colour and simplistic style; especially, how to use 2D geometrical form to create 3D space. Thus, first of all, in my future work or study, the key is to develop the research of solid shape and negative space in the application of fashion design. This is because I have always been interested in the concept of space between the human body and garment. Therefore, I will develop this concept further and design a collection of garment to explore this idea.

In the meanwhile, I have realised that I have a great deal of curiosity about the balance of proportions and details in Abstract Art (Quinn, 2003; Walker, 2011; Dimant, 2010). I believe that studying and researching this area will also help a designer to design ready-to-wear garment. What is more, I will study more on other topics I am really interested, like how to use the balance of structure in constructivism abstraction in fashion; how to get inspiration from Mondrian's painting proportion detail and use it in garment's overall design, as well as how to use modern mind to transfer abstract Geometric style to fashion garment language. For example, the artists of "Constructivism" abstraction show their idea by reducing some unnecessary ingredients supported by structural rheology, human body mechanics and geometry. I think this idea can be used in fashion area, maybe I can analyse the figure of garments by solid geometry, quantification of volume, and other analysis to change the main structure of the garment, just like the red and black boiled wool jacket and skirt in Yohji Yamamoto Autumn/winter 2009 (Salazar & Yammoto, 2011).

Finally, I find myself continually thinking about how to combine my research with Chinese traditional garment together, in order to create a new concept of fashion. This idea is from Japanese designers: Rei Kawakubo, Yohji Yamamoto and Issey Miyake. The

最后，不断思考着如何将所研究的内容跟中国的传统时装设计结合在一起，创造出新的时尚概念。这个概念的产生也是通过对日本设计师的设计理念研究得出的，如川久保玲，山本耀司、三宅一生等，将日本和服和"禅意"与现代服饰相结合，诞生了属于日本的美感服装。

总而言之，通过这些设计与研究，为中国的时装事业做出一点贡献。本人会深入的探索和发展本人的一些想法，然后一步步实现它。我更希望，通过我的设计和研究，能够启发和引导更多的设计师深入这个领域，进行进一步的创新开发。

common feature of their garment is mixing the heavy traditional garments with Japanese concept of wabi-sabi (incomplete aesthetics), based on which they have developed and created their current modern garments.

In conclusion, I hope I can use my design and my research to help Chinese fashion industry to make a different attempt. I like to develop some of these ideas, and then achieve it step by step. I even hope my design and research could made impact more and more designers to be interested in this area and to co-operate with me in this design area.

参考文献 / Reference

1，Black. S. (2006a) 'Yohji Yamamoto' Fashioning Fabrics. 1st publication. London.

2，Black. S. (2006b) 'Issey Miyake' Fashioning Fabrics. 1st publication. London.

3，Director, A. B. (1982), Barbara Hepworth 1st edn. London: The Tate Gallery

4，Frankel, S. (1997) "Between the Place" Guardian Weekend Magazine, Fukai, A.; Vinken, B.; Frankel, B.; Kurino, H.; Nie, N. (2010a) 'Rei Kawakubo' Future beauty: 30 years of Japanese fashion. Merrell Publishers

5，Fukai, A.; Vinken, B.; Frankel, B.; Kurino, H.; Nie, N. (2010b) 'Yohji yamamoto' Future beauty: 30 years of Japanese fashion. Merrell Publishers Washida, K. (2002), "The Past, the Feminine, the Vain', In Talking to Myself by Yohji Yamamoto. Milan

5，Hepworth, B. (1962) 'The Statement by the Artist' Art Gallery

6，Miyake, I. (2010) Miyake Design studio, Tokyo

7，Quinn. B. (2003) The fashion of Architecture. 1st Publication. London. P2-p11

8，Shimizu S. (2005) Comme des Garcons, Tokyo

9，The Tate Gallery (1982) Barbara Hepworth a guide to the tate gallery collection at london and ST Ives, cornwall The Tate Gallery Publications Department

10，Thistlewood, T. & MacPhee, A. (1996) Barbara Hepworth reconsidered Liverpool: Liverpool University Press and Tate Gallery Liverpool

11，Thistlewood, D. & MacPhee, A. (1996) Barbara Hepworth reconsidered Liverpool & London: Liverpool University Press and Tate Gallery Liverpool

参阅书目
Bibliography

1，Hamilton, G. H. (1976) "Picasso, Pablo Ruiz Y". In W. D. Halsey. Collier's Encyclopedia. 19. New York: Macmillan Educational Corporation. pp. 25–26.

2，Boddy-Evans, M. (2009) Abstract Art -- What Is Abstract Art or Abstract Painting, retrieved January 7, 2009

3，Themes in American Art - Abstraction, retrieved January 7, 2009.[online] Available at http://www.nga.gov/education/american/abstract.shtm [Accessed on 28/04/2011]

4，Read, H. (1959) A Concise History of Modern Art, London: Thames and Hudson

5，Grohmann, W.; Kandinsky, W. (1958) Life and Work. New York: Harry N Abrams Inc.,

6，Kandinsky, W. (1965) Point and Line to Plane. New York: Dover Publications

7，Kandinsky, W.; Sadler, M. T. (2001). 'Concerning the Spiritual in Art'. Edited by Adrian Glew New York: MFA Publications and London: Tate Publishing

8，Blavatsky, H. P. (1889). The Key to Theosophy. London: The Theosophical Publishing Company. Theosophy Trust Books. 2007

9，Blavatsky, H. P. (1977). The secret doctrine : the synthesis of science, religion, and philosophy. Theosophical University Press.

10，Chadaga, B. (2000) "Art, Technology, and Modernity in Russia and Eastern Europe" Columbia University Model for `Constructed Torso' 1917, reassembled 1981 Tate Gallery http://www.tate.org.uk/servlet/ViewWork?workid=21293

11，Gabo, N.; Naum, C. (1987) Sixty years of constructivism Tate Gallery Publications,

12，Whitford, F. (1987) Understanding abstract art, Barrie & Jenkins, 1987

13 , Yve- Alain B.; Joosten, J.; Rudenstine, A. Z.; Janssen (1995) Hans Piet Mondrian, Canada: Litter Brown and Company,

14 , Dabrowski, M. (1985) Contrasts of form: Geometric Abstract Art1910-1980, Museum of Modern Art

15 , Mondrian, P.; Riley, B.; (1997) Gemeentemuseum, H. Mondrian: Nauture to Abstraction. Tate Gallery Pub

Attachment
附件

附件一
Attachment One

读研毕业作品在英国 Art and Design gallery 展出
"The Imagination of space "exhibition in Art and Design Gallery

设计师与参观者交流
Designer discussed with visitors

设计师与艺术家交流
Designer discussed with artists

附件二
Attachment Two

Iris Van Herpen 公司简介及邀请加入设计团队函件

Iris Van Herpen 品牌是由荷兰新锐女性设计师 Iris Van Herpen 于 2007 年创立的同名服装品牌。是荷兰首屈一指的一线高级定制服装品牌。近几年，这个品牌的巴黎高级定制服装作品可与 Chanel，Dior，Balenciaga 等老牌高级定制相抗衡，被国外时尚界称之为————高不可攀的艺术品。

品牌创始人 Iris Van Herpen 是一位年轻且才华横溢的女性设计师，毕业于荷兰艺术学院，十分擅长从服装本身的材质着手，将工艺，手工和最新科技完美的相结合，并且辅以夸张的造型。她的高级定制服装设计在巴黎高级定制秀中大受好评，同时囊括了荷兰设计界最高奖项"荷兰设计时尚大奖""配件设计大奖"和"2010 年荷兰设计师"大奖。

2011 年，我将我的简历及作品登在英国设计师招聘网上，Herpen 女士看了我的作品非常的欣赏，认可我的设计，对我发出邀请。希望我加入他们的设计团队。

但是虽然英国是欧盟国，却并未加入欧元区，不是申根国，所以我无法按时顺利得到赴荷签证。当我向 Iris Van Herpen 公司说明这一情况后，他们告知可以继续等我至 2012 年 3 月份，并由 Herpen 女士的专人助理开出一张设计师助理工作邀请函（见后页）。

非常遗憾，最终还是因为签证原因未能如愿前往。

Yue Xu is invited to do a designer assistant at IRIS VAN HERPEN, starting the 12th of March 2012 and ending the 13th of July 2012.

Arnhem, 10th of November 2011

Sarah Blom
Assistant Iris van Herpen

附件三
Attachment Three

《空间的遐想》获2012年第六届创意中国·全国设计艺术大奖赛 服装设计类 一等奖，并已入选《中国创意设计年鉴·2012》

　　此次创意中国设计大奖是由中国设计师协会与成都蓉城美术馆联合主办，作为中国创意设计界最高规格的设计大奖，旨在展现中国创意设计领域的最新成果，推进我国设计人才的知名化，国际化进程。大赛注重设计与艺术的结合，坚持交叉与融合的标准，引入国际设计专家参加评选，代表着中国当前设计研究与创作的年度水准，以非凡的创作力推动中国设计走向世界。

————摘引自创意中国·第六届全国设计艺术大奖赛组委会《获奖喜报——暨〈中国创意设计年鉴〉入编通知》